D1200454

Perennial Weeds

Perennial Weeds

Characteristics and Identification of Selected Herbaceous Species

Wood Powell Anderson

Iowa State University Press
Ames

Wood Powell Anderson, Professor Emeritus of Agronomy, New Mexico State University, worked jointly in the Department of Agronomy and the Agricultural Experimental Station.

Design by Dennis Anderson

Iowa State University Press
2121 South State Avenue, Ames, Iowa 50014

Orders: 1-800-862-6657
Office: 1-515-292-0140
Fax: 1-515-292-3348
Web site: www.isupress.edu

First edition, 1999

Library of Congress Cataloging-in-Publication Data

Anderson, Wood Powell
Perennial weeds : characteristics and identification of selected herbaceous species / Wood Powell Anderson.—1st ed.
p. cm.
Includes bibliographical references and index.
ISBN 0-8138-2520-2
1. Weeds—Identification. 2. Weeds—Control. I. Title.
SB611.A5 1999
632'.5—dc21 99-35933

The last digit is the print number: 9 8 7 6 5 4 3 2 1

Contents

Preface

Most perennial weeds propagate and spread by seeds, but they differ in how they propagate and spread vegetatively. Knowledge of these variations is vital to understanding why some weeds are perennial, and why they grow year after year in spite of persistent control efforts.

The first chapter covers the general characteristics of perennial weeds, distinguishing between simple and creeping perennials and describing how they propagate, spread, and overwinter. The second chapter presents the general means by which perennial weeds are controlled. The remaining chapters cover the characteristics and identification of 28 herbaceous perennial weeds—grouped as grass, grasslike, and broadleaf perennials. The broadleaf perennials are subgrouped according to their vegetative propagation and spread. The selected weeds exhibit characteristics by which herbaceous perennial weeds, in general, differ. Each is a problem weed in one or more agricultural regions of the United States and Canada, except for stinging nettle, a nuisance weed.

The distribution maps give only a general approximation of the area in which a specific plant may be considered a weed. It is recognized that local infestations of a particular weed species occur that are not included on the weed distribution maps; such infestations are ongoing occurrences as weed species are introduced into new areas. Also, due to variations in elevation, climate, and soil conditions, a particular weed may not occur in every location within the geographic area shown in its respective distribution map. The denser cross-hatching on the maps indicates the area where infestations of the particular weed is of greatest economic importance.

This publication is especially directed to those persons concerned with understanding, teaching, identifying, and controlling herbaceous perennial weeds. Professors and instructors at universities and community colleges teaching courses involving weeds and weed control will find it especially helpful as a supplemental text for use in their courses. It should also be useful to agricultural chemical dealers, farmers, golf course and park superintendents, and homeowners, aiding them in better understanding perennial weeds and why they are difficult to control.

I wish to thank Dr. Arnold P. Appleby, Oregon State University, Professor Emeritus, Department of Crop and Soil Science, for his careful review of the manuscript and helpful suggestions for its improvement, and to express appreciation to Dr. Donald J. Cotter, New Mexico State University, Professor Emeritus, Department of Horticulture, for his helpful suggestions in the early stage of developing the manuscript.

Part One

Introduction

1

Characteristics of Perennial Weeds

Introduction

Perennial weeds are distinguished from annual weeds in that they live 3 or more years. There are *terrestrial* and *aquatic* perennial weeds. Terrestrial perennial weeds may be grouped as *herbaceous* or *woody.* Herbaceous perennials have succulent, nonwoody, aboveground vegetation that is killed by severe drought, frost, and below-freezing temperatures (e.g., horsenettle and leafy spurge). Herbaceous stems generally undergo little or no secondary growth; that is, no increase in diameter due to cambial activity. Woody perennials have woody, aboveground stems that survive most inclement weather conditions, even though they may defoliate during adverse growing conditions (e.g., mesquite and sagebrush). Woody stems are characterized by secondary growth occurring during the 1st, 2nd, and later years of growth. This secondary growth is evident by an increase in stem diameter resulting from new cells laid down by the activity of the vascular and cork cambiums. Grass plants are primarily herbaceous, whereas those of broadleaved plants (dicots) may be either herbaceous or woody. The author has chosen to confine the limits of this publication to terrestrial, herbaceous, perennial weeds.

Herbaceous perennials are grouped as *simple perennials* and *creeping perennials.* Simple perennials reproduce primarily by seeds and by root crown buds. The root crowns are supported by fleshy taproots that produce new plants year after year, sustained by stored food reserves in the enlarged taproot (e.g., dandelion, curly dock). Simple perennials may have other types of noncreeping roots, such as tuberous roots, woody taproots, and fibrous roots, or they may grow in clumps or tufts with a fibrous root system. The plantains, broadleaf and buckhorn, vary from this in that the long, thin taproot is replaced later with a short rhizome and a bundle of fibrous roots. Simple perennial weeds depend on seed dispersal to invade near and distant areas. Examples of simple perennial weeds and their characteristic root systems are given in Table 1 of the Appendix.

Creeping perennials are distinguished by having creeping roots (e.g., field bindweed), rhizomes (e.g., johnsongrass), rhizomes and stolons (e.g., bermudagrass), stolons (creeping woodsorrel) and/or aerial runners (e.g., ground ivy) that elongate and produce new plants from reproductive buds on these organs (Tables 2, 3, and 4 in the Appendix).

PROPAGATION

Perennial weeds propagate by *sexual* (seeds) and/or *asexual* (vegetative) means. Some perennial weed species are capable of prolific seed production; for example, curly dock and johnsongrass may produce 30,000 and 80,000 seeds/plant, respectively. In contrast, the seed production of some other perennial weed species is sparse; e.g., purple nutsedge, Russian knapweed, and wild garlic.

Asexual reproduction results from new plants arising from buds on vegetative reproductive plant parts. Examples of asexual reproductive plant parts include vertical and horizontal roots; creeping rhizomes; noncreeping short rhizomes; creeping stolons; aerial runners; decumbent stems rooting at nodes in contact with the soil, but not true stolons; bulbs and/or aerial bulblets; and tubers. Examples of each of these types of perennial weeds are given in the Appendix, Tables 2, 3, and 4. Many perennial weeds can also propagate from reproductive buds located along their underground vertical stems. Perennial weed seedlings cannot reproduce asexually until they have developed a vegetative reproductive organ, such as a root or a rhizome.

Perennial weed species have characteristic means of asexual propagation. For example, dandelion and yellow rocket propagate asexually by axillary crown buds at apex of fleshy taproots; chicory by axillary crown buds at apex of woody taproots; Canada thistle, field bindweed, and leafy spurge by adventitious buds on creeping roots; johnsongrass, quackgrass, stinging nettle, and yellow woodsorrel by axillary buds on creeping rhizomes; common bermudagrass by axillary buds on both rhizomes and stolons; creeping woodsorrel by axillary buds on stolons; field chickweed by axillary buds on decumbent stems; creeping buttercup and healall by axillary buds on runners; purple nutsedge and yellow nutsedge by axillary buds on tubers; and wild garlic by bulbs and aerial bulblets.

The asexual reproductive parts of many perennial weeds are located in the upper 1- to 4-inch (2.5- to 10-cm) soil layer. At this depth, they are subjected to the tumbling, cutting, tearing, and dragging action of field cultivators. In contrast, the asexual reproductive parts of other perennial weeds, such as Canada thistle, field bindweed, and leafy spurge, extend well below a soil depth of 8 inches (20 cm), a depth generally considered the plowshare depth.

SPREAD

Simple perennials have no natural means of spreading other than by seeds. In some instances, pieces of their root crowns or fleshy taproots may be broken or cut off and carried to new locations. In contrast, creeping perennials spread by seed, and underground horizontal roots or rhizomes, aboveground runners and/or stolons, and by decumbent stems that form new plants at nodes when in contact with moist soil. Examples of simple perennial weeds that spread by seeds are given in the Appendix in Table 1. Examples of perennial weeds that spread by seed and by various vegetative means are given in the Appendix in Tables 2, 3, and 4. Common bermudagrass is an example of a perennial weed that spreads by seed, rhizomes, and stolons. Purple nutsedge spreads by rhizomes (rarely by seed).

PERENNATING ORGANS

The term *perennate* denotes living from season to season. Perennating organs are the vegetative reproductive parts (asexual propagules) of perennial plants, such as fleshy taproots, roots (those capable of developing reproduction buds), rhizomes, tubers, bulbs and aerial bulbils. Perennating organs enable a plant to survive from season to season, living through unfavorable climatic conditions by lowered metabolic activity and resuming high metabolic activity when conditions are again favorable for growth.

Horizontal Roots

Creeping *horizontal roots* are true roots (not rhizomes); characteristically irregular in growth, lack nodes and leaf structures, and give rise to adventitious aerial shoots and adventitious lateral roots at any place along their length. In general, horizontal roots grow at deeper depths in soil than do rhizomes. They may turn downward at any point and penetrate to greater depths in the soil as thick, vertical roots. Aerial shoots commonly develop from buds located at random along horizontal roots, and they commonly arise from buds located just above the bend as the horizontal root turns downward; in contrast, lateral, roots normally arise from buds located just below the bend. Perennial weeds that spread by creeping horizontal roots do not also spread by creeping rhizomes.

Rhizomes

Rhizomes are, by definition, creeping, horizontal, underground stems. Some rhizomes may be short, blunt, and noncreeping, but still horizontal. Rhizomes have nodes, internodes, axillary buds, and scalelike leaves at the nodes. Creeping rhizomes normally grow at shallower depths than do creeping roots, and they may turn upward at any point, but not downward. Aerial shoots, secondary rhizome branches, and fibrous roots only arise from axillary buds at the nodes of rhizomes. Perennial weeds that spread by creeping rhizomes do not also spread by creeping horizontal roots. Sometimes, in the scientific literature, underground vertical stems, capable of propagation, are referred to as "vertical rhizomes." This is incorrect and the practice should be discouraged.

Tubers

A *tuber* is the enlarged tip of a rhizome (e.g., yellow nutsedge and potato). A mature tuber has several axillary buds ("eyes") capable of sprouting to form aerial shoots or new rhizomes. Tubers are perennating organs that serve for reproduction and food storage.

Bulbs

A *bulb* is a large, rather globose, bud with a small basal stem at its lower end from which grow fleshy, scalelike, overlapping leaves (e.g., onion, wild onion, wild garlic). It consists largely of storage leaves. Bulbs are perennating organs that serve for reproduction and food storage.

Modified Stems

Stems may be greatly modified from a basically upright cylindrical structure. Stem modification can be divided into belowground forms (rhizomes, tubers, corms, bulbs) and aboveground forms (crowns, stolons, runners). Many of these stem modifications contain large amounts of stored food, and they are especially significant in propagation.

Rhizomes are horizontal underground stems, and they may be compressed, fleshy, thick, or slender, with elongated internodes. Tubers are greatly enlarged portions of rhizomes, and they are typically noncylindrical. Corms are compressed, fleshy, underground stems having few nodes. Bulbs are short, flattened, or disc-shaped stems surrounded by fleshy leaflike structures called scales. They may enclose shoots or flower buds. Bulbs commonly grow underground or at ground level. Bulblike structures (bulbils) may be formed on aerial stems. Bulbs and corms are found only in some monocotyledonous plants, such as wild garlic.

There are two types of crowns, *root crowns* and *stem crowns*. Root crowns form at the top of fleshy taproots, as in dandelion or the garden carrot; they are compressed stems. Stem crowns are portions of stems that form buds and new shoots following death of their aboveground parts, as by cutting or adverse weather conditions. In general, root and stem crowns are located just below or just above ground level. Roots, aerial shoots, leaves, or flowers may arise from axillary buds at the crown nodes.

Stolons and *runners* are similar but different. They are similar in being aboveground, horizontal, creeping stems, but they differ in how their creeping sections (internode chains) originate. The continuous chain of internodes of a stolon arise from the terminal bud of the internode last formed (termed *monopodial*), as with bermudagrass and creeping buttercup. If, as sometimes happens, the terminal bud produces a flower, then formation of another internode from the terminal bud is no longer possible; any subsequent stem growth is due to branching (i.e., formation of an apparent main axis from successive secondary axes [termed *sympodial*]).

Runners are composed of chains of internodes arising from an axillary leaf bud near the end of the previously formed internode (sympodial). For example, runners of silverweed cinquefoil (*Potentilla anserins*) produce a pair of opposite leaves at the apex of a newly formed internode. A bud in the axil of one of these leaves gives rise to the next internode of the runner. If the runner roots at this node, the bud in the axil of the other leaf forms a rosette, and eventually a new plant. Normally, the terminal bud of a runner will not develop further, or it may give rise to a flower. Thus, runners are composed of a chain of connected internodes arising from an axillary bud at the end of the previously formed internode. The strawberry (*Fragaria vesca*) plant is often used as an example of a herbaceous perennial that reproduces vegetatively by runners.

PERENNIAL WEED TYPES

Table 5 of the Appendix groups lists selected perennial weeds as to type.

LITERATURE INACCURACIES

It is not unusual to read in weed identification guides that a given plant *propagates* by roots, rhizomes, or creeping rootstocks. This is inaccurate and misleading. These perennating structures are responsible for the *spread,* not the propagation, of perennial plants. At best, such references can be accepted as a generality that assumes the reader understands that the buds (adventitious or axillary) on these plant parts are actually responsible for asexual propagation.

The term *rootstock* is often encountered in the scientific literature to denote a perennating structure, without revealing whether it is a root or rhizome. Thus, the reader is left in a quandary as to precisely which type of perennating structure is involved. Due to its ambiguity, the term should be avoided or used with caution. If used, the term *rootstock* should be clearly defined for the reader.

At times, a root is erroneously identified as a rhizome, or vice versa. This error can be corrected by examination of the organ involved. *Roots* do not have nodes or internodes, and their branches arise from the *pericycle* [a layer of cells lining the interior of the endodermis and surrounding the vascular tissue (xylem, phloem) within the root]. Microscopic examination of a cross section will show the characteristic cellular arrangement of a root. Rhizomes have nodes, internodes, axillary buds, and scalelike leaves at the nodes. The axillary buds arise from dedifferentiated cortical tissue, a broad cylinder of thin-walled living cells (*parenchyma*) located between the endodermis and epidermis of the root. Rhizomes may be slender (e.g., bermudagrass), relatively large and fleshy (e.g., johnsongrass), or woody (e.g., western ironweed). Microscopic examination of a cross section of a rhizome shows the cellular arrangement characteristic of stems.

WINTER KILL AND DROUGHT

The aerial vegetative growth of herbaceous perennial weeds is commonly killed to or just below the soil surface by freezing or near freezing temperatures. Examples of such herbaceous perennial weeds include bermudagrass, johnsongrass, quackgrass, Canada thistle, field bindweed, horsenettle, leafy spurge, and silverleaf nightshade. Severe drought during the growing season also may kill the aerial vegetation of perennials to ground level. When climatic conditions are again favorable for growth, the plants will regenerate new aerial shoots from buds on their perennating structures. In general, succulent belowground roots and rhizomes may be killed when the temperature of the soil in which they are embedded falls below freezing.

DORMANCY

Dormancy in perennial weed propagules is a survival mechanism, enabling the species to survive periods of adverse growing conditions. Seeds of many perennial weed species can remain dormant for many years, as much as 20 years in the case of field bindweed.

The inability of some weed species to make vegetative growth during part of the growing season is of considerable importance in the timing and effectiveness of weed control measures. Three types of dormancy: *internal, late spring,* and *summer* and two means of vegetative growth inhibition: *correlative inhibition* and *apical dominance,* are discussed below.

Internal Dormancy

Perennial weed plants become dormant at the end of their respective growing season, usually with the onset of freezing temperatures and/or short days (long nights). Although the aboveground portions of the plants may be killed by frosts and freezing temperatures, the belowground portions are generally protected from such adverse conditions and remain alive and capable of regeneration. During this period the plants are exhibiting a form of dormancy termed *internal dormancy.* Internal dormancy, a natural phenomenon, allows the underground perennating organs to survive adverse winter conditions, even though the aerial vegetation is dead. The plants regenerate from buds sprouting on the perennating structures after internal dormancy ends naturally, usually in the spring. Some perennial weed species (e.g., western ironweed) exhibit internal dormancy during the growing season, even though their aerial parts are not dead, and resume growth later in the season.

Late Spring Dormancy

Some perennial weeds (e.g., quackgrass) exhibit a type of dormancy termed *late spring dormancy.* In this type of dormancy the axillary buds on the rhizomes remain inactive, even if the rhizomes are segmented and apical dominance is no longer a factor. This type of dormancy occurs in late spring under conditions of high moisture, cool temperatures, rapid top growth, and rapid development of new rhizomes.

Summer Dormancy

Many perennial weeds become dormant during hot, dry periods in the growing season, even though moisture is not a limiting factor. This inactivity is associated with high temperatures and long days (12 h or more of light). This type of dormancy is termed *summer dormancy.*

Correlative Inhibition

Correlative inhibition is the inhibiting influence on the growth of a bud or other organ by another part of the plant.

Apical dominance is the correlative inhibition of axillary buds by the apex (tip) of the parent organ.

During a single growing season, at one time or another, some perennial weeds (e.g., quackgrass, leafy spurge, and western ironweed) exhibit internal dormancy, immaturity (basal buds), summer dormancy, correlative inhibition, and apical dominance relative to bud development on rhizomes and shoots.

2

Control of Perennial Weeds

Introduction

Perennial weeds are not a problem if they do not produce seeds, nor spread vegetatively, nor reproduce asexually, nor successfully compete with desired plants, nor be poisonous to humans and/or their animals, nor impair one's view and/or aesthetic values. The first step in deciding how to solve a perennial weed problem is to identify correctly the weed species and understand its characteristic growth pattern.

Practices used to control perennial weeds are the same as those used against other kinds of weeds. They may be grouped as *preventive, cultural, mechanical* (*physical*), *chemical,* and *biological.* For best control, a combination of two or more practices is often needed. For example, a combination of prevention, cultural, and chemical, or prevention, mechanical, and chemical, or mechanical and biological. One must be aware that a control practice that is effective in one locale may not be effective in another due to differences in the weed species involved, weed biotypes, soil type, climate, and cropping systems.

The term *control* does not necessarily mean eradication of a weed species from a given area. Control infers population reduction and competitive suppression of the weed to a level that is aesthetically and economically acceptable. Long-term economic and cultural benefits can be obtained by minimizing reproduction and spread of the weed. The aesthetic value of weed control is often ignored when tabulating the cost of perennial weed control.

Once established, perennial weeds are the most difficult group of weeds to effectively control, usually requiring years of persistent effort. *Prevention* is the surest method of control; i.e., preventing the introduction of weed seeds or their vegetative propagules into a field or region. Perhaps the next best method of control is to kill the weed seedlings shortly after emergence and before establishment, usually in the 1st week or 2 after soil emergence. Seedlings of perennial weeds are as easy to kill as are seedlings of annual and biennial weeds.

The difficulty in controlling perennial weeds lies with their long-lived seeds (sexual) and persistent vegetative (asexual) propagules. Seeds of some perennial weeds live and persist in the soil for 40 or more years. Vegetative reproductive parts of many undisturbed perennial weeds live in the soil for more than 10 years, and some as long as 50 years. For ideal control, these sexual and asexual propagules must be killed before they mature to prevent new populations of the weed species from increasing and becoming established.

Perennial weeds manufacture energy-rich carbohydrates by photosynthesis in their aboveground, green vegetation. The carbohydrates are subsequently translocated to, and stored in, the underground plant parts, such as roots, rhizomes, tubers, and bulbs. This food reserve is used to support growth of these underground organs and the development of aerial shoots.

Preventing photosynthesis by killing the aboveground vegetation by tillage or contact herbicides will eventually exhaust their stored food supply, as they must repeatedly draw on this food supply to produce new aerial shoots to produce more photosynthates. In time, this practice will result in death by starvation of belowground plant parts. To be successful, it is necessary to kill the original aerial shoots and any regrowth that occurs later. Thus, such a control program requires persistent effort over a period of 3 or more years, depending on the weed species and environmental conditions involved.

Understanding the dormancy, reproduction, and growth cycles of perennial weeds is significant to effective control. Dormancy in seeds and/or asexual parts delineates the seasonal growth of perennial weeds and minimizes the effectiveness of a single control measure. Effective control is based on the knowledge that dormancy will end, and that seed germination and/or asexual regeneration will follow.

A major obstacle to the control of creeping perennial weeds is their capacity to regenerate from buds on roots, rhizomes, stolons, tubers, or other specialized organs of perennation. When such weeds are established and growing undisturbed, only a small proportion of these buds are involved in producing new growth and aerial shoots. The sustained inactivity of these buds is attributed to *correlative inhibition* associated with the aerial shoots (i.e., the inhibiting effect of the aerial shoots on the growth of the root buds). However, many of the dormant root buds will be induced to grow if the aerial shoots are injured or killed, usually resulting in a flush of many new aerial shoots. Thus, any control practice, such as tillage or mowing, that can eliminate or significantly reduce the correlative inhibition of bud growth would make the affected weeds more susceptible to subsequent control measures. Such practices would induce bud growth and the subsequent emergence of new aerial shoots which, in turn, would reduce stored food reserves and expose the aerial shoots to destruction by cultivation and/or interception of foliar-applied herbicides.

PREVENTION

Prevention as a weed-control practice has the objective of preventing the introduction and spread of one or more weed species into a given area. The most important preventive measure is to plant weed-free crop seeds and to avoid using hay harvested from weed-infested fields. When necessary, laws may be enacted by legislation at the national, regional, state, and/or local level in an effort to control the spread of especially noxious perennial weeds. However, such laws are generally hard to enforce, and they may result in bias in enforcement and pit neighbor against neighbor.

Preventive control measures are often circumvented by the spread of weed seeds and other propagules by wind, flowing water, animals, and the various enterprises of people.

MOWING AND GRAZING

Perennial weeds are generally mowed to prevent seed production, to starve underground plant parts, and for aesthetic reasons. To be effective, mowing must be done before viable seed are formed, and frequent mowings during the growing season may be required over several years to deplete the stored food reserves. However, it is not unusual for mowed weeds to produce flowers and, subsequently, seeds on vegetation located below the cutter bar.

Grazing may be as effective as mowing in preventing seed production and spread of perennial weeds, but the grazing animals must find the weed plants palatable. Otherwise, they may graze all around the clusters of perennial weeds, thereby removing plant competition and leaving the weeds to thrive and spread. Also, weed propagules may be spread if they become attached to the hooves and hair of the grazing animals.

Grazing and mowing commonly result in increased populations arising in the perennial weed colonies following the release of dormancy in buds previously held in check by intact aerial shoots (correlative inhibition). This can aid in depleting stored food reserves if the new aerial growth is not allowed to flourish. Usually, mowing and/or grazing perennial weeds is not effective in eliminating perennial weed infestations.

SMOTHER CROPS

Smother crops are crop plants that grow rapidly and form dense stands that shade and crowd out other plant species. Highly effective smother crops used to combat perennial weeds include fall-planted wheat and rye, summer-planted sorghum, and Sudan grass planted for hay. During the fallow period between crops, the land should be kept weed-free by repeated clean-cultivations or with herbicides to prevent new aerial shoots from photosynthesizing and replenishing food reserves in underground storage organs. Row crops, especially those grown in wide rows, and low-growing crops do not make good smother crops. Perennial weed control may be enhanced by combining the use of smother crops with herbicide applications.

Alfalfa is perhaps the most effective smother crop used to control perennial weeds. Alfalfa has three advantages as a smother crop: it forms a dense stand; the planting normally stays undisturbed for 3 or more years; and, depending on location, the crop is harvested three to six or seven times a year. Unfortunately, alfalfa is commonly dormant during the winter months.

Mulches are effective in controlling perennial weeds if the aerial shoots do not grow through the mulch to light. Black plastic sheeting placed to cover the soil surface is effective as a control measure if retained in place free of tears or punctures. To be effective against perennial weeds, mulches or black plastic must be kept in place, and light-tight, for 3 or more years. Shoots of some perennial weeds, such as the nutsedges, easily penetrate plastic sheeting.

CULTIVATION

Clean-cultivation, resulting from multiple cultivations at intervals of 2 to 3 weeks each year during the growing season for 3 or more years, has been effective in controlling many perennial weed species. Unfortunately, such a practice is often impractical, as the land must be out of production during this time. A single clean-cultivation during the growing season will provide only temporary control of perennial weeds, and regrowth may actually result in a denser stand of aerial shoots than were present prior to the cultivation.

The optimum, practical depth of cultivation for control of perennial weeds is 4 inches (10 cm). The advantages of deeper tillage are offset by the increased power required. The control of seedlings and plants less than 1 year old is relatively easy with shallower cultivation, but it is much more difficult to kill older established plants because of their vigorous regrowth from underground propagules.

Intensive cultivation reduces the plant's stored food reserves, reduces root and rhizome quantities, segments roots and rhizomes, tumbles and disrupts underground propagules, and severs new shoots as they develop. It kills seedlings before they develop perennating structures. Cultivation is most effective when performed with sweep-type implements (e.g., duckfoot sweeps) that cut the shoots from the roots or rhizomes. A serious consequence of cultivation is the cutting, dragging, and relocation of the asexual propagules of perennial weeds within a field where they will sprout and start new colonies. Seedlings of perennial weeds are most susceptible to control by cultivation when they are 1 to 2 inches (2.5 to 5 cm) tall.

With perennial weeds that propagate from horizontal and vertical roots positioned in the soil below the depth of tillage, these deeper propagules will continue to grow and produce aerial shoots until their food reserves are exhausted, even though their shallower parts are injured or killed by the tillage tools.

CHEMICAL CONTROL

Control of perennial weeds is greatly enhanced by the use of appropriate soil- and foliar-applied herbicides. Effective soil-applied herbicides kill underground plant parts upon contact, or they are absorbed by underground parts and translocated in phytotoxic amounts from the point of absorption to other plant parts, especially to regions where rapid cell division is taking place, such as the buds.

The most effective foliar-applied herbicides are ones that translocate from the foliage throughout the underground vegetative network in phytotoxic amounts that kill these parts, especially the buds (e.g., amitrole applied to poison ivy). Some foliar-applied herbicides do not translocate and do no more than kill the aboveground vegetation of perennial plants (e.g., bromoxynil, diquat, paraquat), and regrowth soon appears. Other foliar-applied herbicides translocate into the shallow network of underground parts, but they leave undisturbed the network of parts located more deeply in the soil (e.g., 2,4-D or glyphosate applied to leafy spurge).

It is essential to select appropriate herbicides for use against specific perennial weed species. Such information is available from state agricultural extension services, local agricultural chemical dealers, and commercial companies marketing herbicides.

In general, the best time to apply herbicides to the foliage of established perennial weeds is when they are in the rosette stage, in the bud to full-bloom stage, or in the fall before growth is stopped by frost. Seedlings of perennial weeds are most susceptible to control with herbicides when 1 to 4 inches (2.5 to 10 cm) tall.

It is rare for one application of a herbicide to kill all propagules of a perennial weed, and if all are not killed, vegetative regrowth will occur in the year of treatment or in the following growing season. Two examples, among many, of effective control of perennial weeds with foliar-applied herbicides are the excellent control of stinging nettle with esters of 2,4-D and of poison ivy and poison oak with amitrole.

BIOLOGICAL CONTROL

Biological control uses living organisms to reduce or suppress the population of targeted weeds to an acceptable level. To date, the principal biological agents have been pathogenic fungi, phytophagous (plant-eating) insects, herbivorous (plant-eating) fish, and goats and sheep. Perennial weeds are the most logical targets for biological control measures. However, the use of biological agents for weed control is still in its infancy.

Leafy spurge infestations often occur on lands unsuitable for cultivation, and many areas so infested cannot be traversed by conventional spray equipment. Thus, leafy spurge is a prime candidate for the use of biological control agents. However, more time and research are needed for advancements to occur in the biological control of leafy spurge.

Goats and sheep have long been used for weed control. Sheep

mostly graze flowering plants (forbs), while goats generally feed on shrubs. Sheep and goats became popular biological control agents to control leafy spurge in the northern Great Plains region in the mid-1980s and the 1990s, complementing the use of herbicides. Leafy spurge is very nutritious and provides good forage for lambs, lactating ewes, and/or lactating nannies. Cattle do not utilize leafy spurge and avoid leafy spurge–infested areas.

Various phytophagous insects have been released in selected leafy spurge–infested areas. One such insect, the minute spurge flea beetle (*Aphthona abdominalis*), was released in 1993. The larvae of this insect feed on young roots, root buds, and root hairs. When under heavy attack by the larvae, the leafy spurge plants are stressed and cannot produce new stems, and root feeding reduces the plant's ability to absorb moisture and nutrients. The adults feed on the flowers and foliage, decreasing sugar production for storage in the roots. There are up to four generations of this insect per year. Because this insect feeds so heavily on leafy spurge plants, it has great potential for reducing leafy spurge populations.

Examples of Control of Perennial Weeds

This section of Chapter 2 is not intended as a "How to control perennial weeds." Control methods are developing so rapidly that recommendations are soon out of date. New effective herbicides are continually being discovered, sometimes replacing older ones. The best control of most perennial weeds is a combination of cultivation, selective herbicides, and, when practical, competitive crops. However, to illustrate the effective control of perennial weeds with cultivation and/or chemicals, selected examples are given below.

Broadleaf Plantain, Buckhorn Plantain, and Blackseed Plantain

Broadleaf plantain, buckhorn plantain, and blackseed plantain are killed by postemergence applications of 2,4-D, 2,4-DP, MCPA, MCPP, dicamba, and various combinations of these herbicides, available as trade named products. These herbicides and combinations are commonly used for control of plantains in lawns and in certain crops. Monocarbamid dihydrogensulfate (Enquik) is a postemergence, nonselective, contact herbicide recommended for plantain control in certain vegetable crops. Enquik is applied as a directed, shielded spray to small, actively growing weeds. Clopyralid + 2,4-D (Curtail, premix) is applied postemergence to the weeds in spring wheat in the early spring. Dichlobenil (Casoron, Norosac) has provided excellent control of weedy plantains in orchards when applied in the fall or early spring while the plants are dormant. Terbacil (Sinbar) is applied in early spring in orchards to control plantains. Metsulfuron (Ally), triclopyr + 2,4-D (Crossbow), and 2,4-D are used in pastures and rangelands for control of weedy plantains.

Creeping Woodsorrel and Yellow Woodsorrel

Seedlings of creeping woodsorrel and yellow woodsorrel are readily removed by handweeding for the first 6 weeks or so after emergence. After this time, rooting stolons of creeping woodsorrel and growth of the rhizomes of yellow woodsorrel make complete removal by handweeding more difficult. Plants may start flowering within 4 weeks after germination. Creeping woodsorrel sets seed most rapidly, and a delay in seedling removal may result in a second flush of offspring from the original plants.

Steam sterilization of soil at 185°F (85°C) for 30 minutes resulted in 97% germination of imbibed seed of creeping woodsorrel; higher temperatures inhibited all germination. Ten-minute pretreatment at any temperature up to 212°F (100°C) still resulted in 100% germination. Thus, steam sterilization of greenhouse soil, which usually involves heating soil to around 180°F (82°C), may not destroy all seeds of *Oxalis* species.

Dithiopyr (Dimension) and isoxaben (Gallery) are labeled for control of both creeping woodsorrel and yellow woodsorrel in turf. Gallery and Snapshot (a premix of isoxaben + oryzalin) are labeled for control of both *Oxalis* species in ornamentals, including container-grown, small fruits, citrus and deciduous fruit orchards, and in noncroplands.

Field Bindweed

The most efficient way to control field bindweed on cropland is to develop a system that combines the use of competitive crops, intensive cultivation, and herbicides. As with other creeping-type perennials, field bindweed control requires diligent effort over a period of years. Not only is it difficult to kill all belowground perennating parts, but long-lived seeds buried in the soil make it almost impossible to eradicate this weed. Field bindweed seedlings will emerge for more than 30 years after an established stand is controlled. Also, lateral roots deep below the soil surface are capable of developing new shoots long after an established stand is thought to have been eliminated. The control of field bindweed is difficult and expensive.

As a rule, herbicides are used in combination with cultivation for control of field bindweed in crop and fallow lands. Greatest control is obtained when herbicide applications are made to actively growing field bindweed in the bud stage, with stems at least 12 inches (30 cm) long. Effective control depends on both absorption of foliar-applied herbicides and translocation to sites of activity in phytotoxic amounts. Herbicide performance often varies due to differences in plant biotypes and environmental conditions. In general, most herbicides are less effective in controlling field bindweed in the western and semiarid plains states than in the wetter, more humid eastern regions of the United States and Canada. Repeat herbicide applications are usually needed for continued effective control of field bindweed.

Competitive Crops

Alfalfa is one of the most competitive crops in suppressing field bindweed. Forage sorghum and sudangrass seeded in narrow rows (solid seeded) about mid-June, after a period of intensive tillage, are excellent competitors in areas where adequate moisture favors crop growth. Field bindweed cannot tolerate shading from tall competitive crops. Row crops, especially grown in wide rows, and low-growing horticultural crops are not effective competitors with field bindweed.

Mulches are effective if the field bindweed shoots do not grow through the mulch to the light. Black plastic is effective if retained in place free of tears or punctures. To be effective, mulches or black plastic must be kept in place, and light-tight, for 3 or more years.

Mowing and overgrazing have not been effective practices in field bindweed control.

Cultivation has been the most common means of field bindweed control, especially in combination with chemicals. The optimum depth for cultivation is 4 inches (10 cm); advantages of deeper tillage are offset by the increased power required. Control of seedlings and 1st-year plants is relatively easy, but it is very difficult to kill older established plants because of vigorous regrowth from buds on the roots and underground vertical stems. Intensive cultivation reduces the root food reserves, reduces root quantity, and severs new shoots as they develop. Just as importantly, it kills seedlings before they develop a perennating root system. Persistent, clean-cultivation for three or more years can eventually kill field bindweed. However, crop production is severely curtailed by this practice. Depletion of the deep root reserves is a long, tedious process.

Cultivation is most effective when performed with sweep-type implements (e.g., duckfoot sweeps) that sever the shoots from the roots. In South Dakota, 95% to 100% elimination of field bindweed in 1 year was achieved by cultivation with overlapping sweeps at a 4-inch (10-cm) depth at 2-week intervals during June and July and at 3-week intervals during August and September. On the Great Plains, 16 to 18 tillage operations, at 2- to 3-week intervals for 3 years or more, are needed to eliminate established stands of field bindweed. However, this method is impractical as it requires at least 2 years of continuous fallow, prevents crop production, and leaves the land open to wind and water erosion. Where alternate winter wheat-fallow farming is practiced on the Great Plains, an effective method is to till at 2- to 3-week intervals during the 15 months of fallow between wheat harvest and seeding.

Effective, long-term control of established plants of field bindweed with foliar-applied herbicides depends on both foliar absorption and translocation of the chemicals in toxic amounts to the extensive root system. Although effective, repeat herbicide applications are usually needed for continuing control. Usually, herbicides are used in combination with cultivation for field bindweed control in crop and fallow lands.

Herbicide performance often varies due to differences in plant and environmental conditions. In general, most herbicides control field bindweed less effectively in the semiarid plains and western states than in the wetter, more humid eastern regions.

Herbicides commonly recommended for control of field bindweed in agronomic crops, in general, are acifluorfen (Blazer), dicamba (Banvel), glyphosate (Jury, Rattler, Roundup), prosulfuron (Peak), trifluralin (Treflan), and premixes of bentazon + atrazine (Laddock S-12), dicamba + atrazine (Marksman), prosulfuron + primisulfuron-methyl (Exceed), and trifluralin (Treflan). Each of these herbicides and premixes, except Marksman, are applied postemergence, over the top, or directly. Marksman is applied preplant or preemergence.

Herbicides recommended for field bindweed control in fallow land include dicamba (Blazer), dicamba + atrazine (Marksman), dicamba + 2,4-D (Weedmaster), and glyphosate + 2,4-D (Landmaster). Each of these herbicides is applied postemergence to field bindweed.

Herbicides recommended for field bindweed control in noncrop areas include dicamba (Banvel); fosamine (Krenite S); hexazinone (Velpar); imazapyr (Arsenal); metsulfuron-methyl (Escort); picloram (Tordon); triclopyr (Garlon 3A); and the premixes 2,4-D + 2,4-DP + MCPA (Dissolve, Par-3, Triamine, Tri-ester). Each of these herbicides is applied postemergence to field bindweed.

Hoary Cress, Lens-podded Whitetop, and Globe-podded Whitetop

Control of these three *Cardaria* species is difficult and requires persistent effort. The extremely persistent reproductive roots, with abundant food reserves, are responsible for the survival and perennial nature of these species. Control by clean cultivation will require 3 consecutive years of intensive tillage to kill the root system of any of these *Cardaria* species. Cultivation should begin early in the spring when the plants are in the bud stage and repeated every 21 days, using a duckfoot cultivator set for 4 inches deep or other blade-type implement. The vertical roots disintegrate from the top down when decay starts following repeated cultivations.

Late-sown crops, such as corn, barley, or beans are effective competitors. Perennial grasses or winter wheat, plus the use of 2,4-D as a selective herbicide, are also effective control measures. Herbicides labeled for control of these *Cardaria* species include 2,4-D and MCPA for use in croplands (e.g., small grains, field corn, peas, rice, sorghum, soybeans, and sugarcane); amitrole-T (Amitrole) for use in woody ornamentals and noncrop areas; metsulfuron-methyl (Escort) for use in pastures and rangelands; chlorsulfuron (Telar) and sulfmeturon (Oust) and 2,4-D for noncroplands. Each of these herbicides are applied postemergence to the *Cardaria* species; Escort, Oust, and Telar are also effective as preemergence treatments.

Applications of 2,4-D at 1.0 lb ae (acid equivalency)/A have been effective in controlling these species in noncroplands. Make the first application in the early bloom stage, the second in July, and the third in September if regrowth is present. Usually 3 years of repeated applications of 2,4-D are needed for complete kill.

Horsenettle and Silverleaf Nightshade

Complete eradication of horsenettle and silverleaf nightshade is difficult. Established horsenettle and silverleaf nightshade stands can be killed by cultivations repeated at 2- to 3-week intervals during the growing season for 3 or more years.

Following such a control program, overlapping sweep-type cultivators set to work the soil 2 to 4 inches (5 to 19 cm) deep are effective tools for this purpose. The objective of such a control program is to deplete the underground plant parts of their stored food reserves. This depletion of stored food is brought about by continually cutting off the aerial shoots, breaking their dominance over the root buds, and forcing these buds to grow and produce new aerial shoots. These new shoots are in turn destroyed by the next cultivation before they can photosynthesize and move these carbohydrates into the underground storage roots, and the cycle is repeated. However, if this control program is followed, the frequent cultivations make the land unavailable for crop production.

There are no herbicides currently available that can be used safely as broadcast treatments in croplands to control horsenettle and silverleaf nightshade, with the exception of dicamba (Banvel) applied early postemergence in field corn. The growth of these two weed species can be suppressed by postemergence applications of primisulfuron (Beacon) in corn and with terbacil (Sinbar) applied to the soil under tree nuts. The premix of triclopyr and 2,4-D (Crossbow) may be used in pastures and rangelands for control of these perennial weeds. Horsenettle and silverleaf nightshade are apparently resistant to applications of 2,4-D and of intermediate resistance to dicamba. Glyphosate (Jury, Rattler, Roundup) may be applied to these weeds as a directed postemergence spray in certain crops, but the crop plants are usually killed if contacted by the spray.

Glyphosate, applied postemergence to plants 6, 9, and 12 weeks after emergence (early bloom, full-bloom to early berry, and late berry, respectively), controlled silverleaf nightshade 90%, 98%, and 85%, respectively. The applications were made as spot treatments using a 3% v/v (liquid volume/liquid volume) of 3 lb ae/gal (10.8 g ae/L) solution of glyphosate. Thus, best control of silverleaf nightshade was obtained when glyphosate was applied to plants in full bloom to early berry, a stage corresponding to 7 weeks after aerial shoots had emerged from the soil.

Picloram applied at 1.1 kg/ha (kilograms/hectare) as a postemergence spray reduced horsenettle stands by 93% during the year of treatment, and the stands were completely eliminated when the treatment was repeated in the two subsequent years. Picloram provides a soil residue that will kill seedlings of these perennial species, but it often proves a hazard to many desired broadleaf plant species.

In noncrop areas, amitrole (Amitrol T), diuron (Direx 4L or 80 DF, and Karmex DF), and the premix triclopyr + 2,4-D (Crossbow) are labeled for control of horsenettle and silverleaf nightshade.

Leafy Spurge

Leafy spurge is extremely difficult to control by chemical means, and it is almost impossible to control by cultural or mechanical means. Infestations of leafy spurge can be controlled in some situations by combinations of timely cultivation, competitive cropping, and herbicide applications.

Leafy spurge is susceptible to the herbicide picloram (Tordon K), but the high cost of Tordon has not justified its use in forage crops, and its soil residues may kill or injure desired vegetation; very small amounts of picloram are toxic to most broadleaf plant species.

A single application of picloram (Tordon K) at 2 lb ae/A (2.24 kg/ha) consistently provided excellent control of leafy spurge plants treated in the bud to early flowering stage. At the end of the third growing season after treatment, control of leafy spurge was 95%, with only slight regrowth apparent. The soil residual of picloram was sufficient to prevent seedling development during this time.

Dicamba (Banvel), at rates of 6 to 8 lb ae/A (6.7 to 9.0 kg/ha), provided good control of leafy spurge when applied to plants in the bud to early flowering stage of growth. In a study in South Dakota, annual spring applications of the butoxyethanol ester of 2,4-D at 1.5 lb ae/A (1.7 kg ae/ha), or a mixture of 2,4-D at 1.0 lb ae/A (1.1 kg ae/ha) + dicamba at 0.5 lb ae/A (0.6 kg ae/ha), + a biannual application of

2,4-D at 0.75 lb ae/A (0.8 kg ae/ha) resulted in satisfactory control of leafy spurge and increased net economic return with minimum economic risk.

Leafy spurge infestations often occur on lands unsuitable for cultivation, and many areas so infested cannot be traversed by conventional spray equipment. Thus, leafy spurge control is a prime candidate for the use of biological control agents. However, more research and time are required for advancements in the biological control of leafy spurge. Various insects have been released in selected leafy spurge infested areas.

One such insect, the minute spurge flea beetle (*Aphthona abdominalis*), was released in 1993. The larvae of this insect feed on young roots, root buds, and root hairs. When under heavy attack by the larvae, the plants are stressed and cannot produce new stems, and root feeding reduces the plants' ability to absorb moisture and nutrients. The adults feed on the flowers and foliage, decreasing sugar production for storage in the roots. Since this species feeds on the foliage, flowers, root hairs, root buds, and young roots, it has great potential for reducing plant densities. The adults are small, measuring about 2.0 mm long and 1.0 mm wide. There are up to four generations of the insect per year.

Some potentially effective insects for control of leafy spurge greatly reduce plant stands at first, but because they do not attack the roots, the plants send up new shoots that are able to produce the sugars for root reserves. Over a long period of time, persistent attacks by the insects may eventually kill the plants.

Despite active efforts to suppress leafy spurge, infestations continue to spread, and the magnitude of the problem continues to increase. In many areas where leafy spurge is a problem, special legislation has been passed to make it illegal to sell hay contaminated with leafy spurge and to allow government authorities to enforce leafy spurge control (e.g., The Wyoming Leafy Spurge Control Act of 1978). Not only are these laws difficult to enforce, they often lead to disagreements between neighbors, biased enforcement, and perhaps inappropriate, costly herbicide applications over large areas.

Purple Nutsedge and Yellow Nutsedge

These two nutsedge species have been among the most difficult weeds to control until recently. In the past few years, the herbicides halosulfuron-methyl (Battalion, Permit, Manage); imazaquin (Image), and imazameth (Cadre) have proven effective for control of both nutsedge species. Battalion is labeled as a preplant incorporated or preemergence treatment in sorghum. Permit is labeled for use in sorghum as a postemergence treatment. Manage is labeled for use in lawns and turf as a postemergence treatment. Cadre is labeled for use in peanuts as a postemergence treatment. The herbicides bentazon (Basagran, Pledge) and MSMA (Ansar, Bueno, Daconate, Drexel MSMA, Setre MSMA) are labeled for postemergence control of yellow nutsedge in lawns and turf and various crops; they do not control purple nutsedge.

Stinging Nettle

Mowing stinging nettle colonies prevents the plants from setting seed, but it increases the plant population by encouraging the growth of numerous bushy shoots. Persistent, repeated ploughing and mechanical cultivation over several years will effectively reduce infestations by eventually killing its extensive rhizome system.

Stinging nettle is very susceptible to the growth-regulator-type herbicides. In noncrop areas, premixes of 2,4-D + 2,4-DP + dicamba (Brushmaster, Cleanout) or 2,4-D + 2,4-DP + MCPP (Dissolve, Par-3, Triamine, Tri-ester) may be applied to stands of stinging nettle as a wetting, foliar-spray. Glufosinate-ammonium [Liberty nc (WS)] may also be applied as a wetting foliar-spray in noncroplands.

Glufosinate-ammonium [Liberty nc (WS)] is labeled for postemergence control of stinging nettle in tree nut orchards. Glufosinate-ammonium (Finale) and glyphosate (Jury, Rattler, Roundup) are labeled for postemergence control of stinging nettle in certain ornamental plantings, and Glufosinate-ammonium (Rely) can be used in certain tree fruit orchards.

Glufosinate-ammonium is a contact, nonsystemic herbicide that suppresses growth of perennial broadleaf weeds. It provides faster knockdown of most emerged weeds than does glyphosate, but it is slower than paraquat. It apparently provides excellent kill of new emerging aerial shoots.

Western Ironweed

The roots, rhizomes, and top-growth of western ironweed are effectively controlled by picloram (Tordon K) applied postemergence at 0.5

lb ae/A or more. However, the relatively high cost of picloram and the toxicity of soil residues of picloram to desired broadleaf plants limit the use of this herbicide for western ironweed control.

Repeated annual postemergence applications of an ester of 2,4-D at 0.5 to 2.0 lb ae/A will control the top-growth of mature western ironweed plants in the year of treatment. Best control is obtained when the 2,4-D ester is applied when the plants are in full-flower.

Mowing is an impractical means of controlling western ironweed. When the dominant apical stem on the rhizome is removed or killed, the vegetative bud next in line sprouts and produces a new aerial shoot.

Part Two

Perennial Grass Weeds Reproducing from Buds on Creeping Rhizomes

3

Johnsongrass
(Sorghum halepense)

Introduction

Johnsongrass (Figure 3.1) is an erect, herbaceous, perennial grass. It is a controversial plant; a noxious weed in many states in the United States and a hay or pasture plant in a few areas in the southeastern United States. Johnsongrass is ranked sixth among the most troublesome weeds in the world. It is listed as a prohibited noxious weed in 19 states of the United States and as a restricted noxious weed in 13 states.

FIGURE 3.1. Johnsongrass. A. Habit, ×0.5; B. spikelet, ×4; C. ligule, ×1.5; D. florets, ×5; E. caryopses, ×5.

Source: McWhorter, C.G. 1973. Johnsongrass as a Weed. Farmers Bulletin No. 1537. Washington, DC: U.S. Department of Agriculture, U.S. Government Printing Office.

Johnsongrass is a serious weed in the major summer crops grown in the United States. It may be found along irrigation ditches, cultivated fields, roadsides, and moist waste places. It grows at elevations of 100 to 6000 feet and flowers from spring to fall. Johnsongrass hybridizes with other sorghum species, resulting in ecotypes that vary in growth habit, morphology, physiology, susceptibility to disease, seed production and germination, and herbicide resistance.

DISTRIBUTION

Johnsongrass is distributed approximately throughout the southern two-thirds of the United States, but occurs as far north as central Washington, southern Michigan, and central New York, Connecticut, and Rhode Island (Figure 3.2). It is native to the Mediterranean region, and was introduced to the United States around 1800 for use as a potential forage crop. By 1840, it was growing in several southeastern states, and by the late 1800s it had been planted throughout much of the United States.

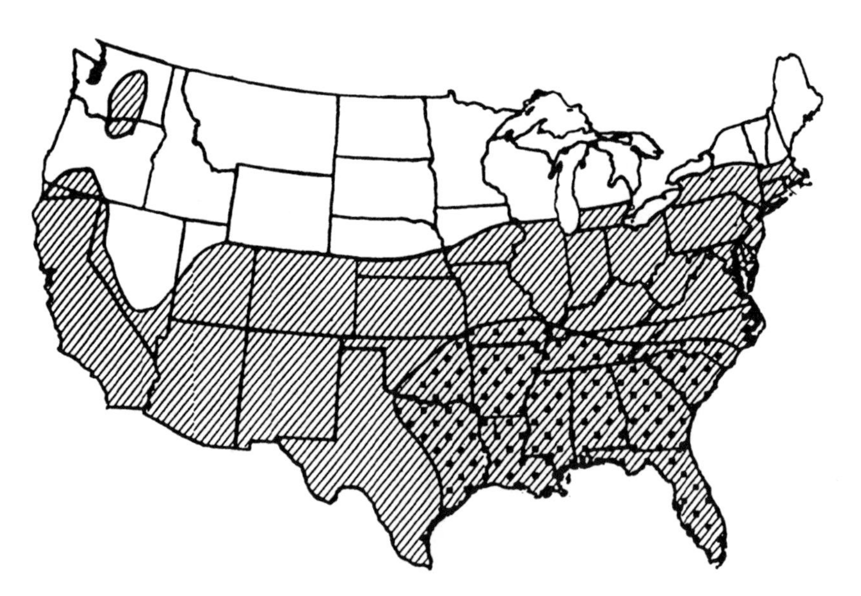

FIGURE 3.2. Distribution of johnsongrass in the United States. Denser cross-hatching denotes infestations of greater economic importance.

Source: Reed, C.F., and R.O. Hughes. 1970. Johnsongrass (*Sorghum halepense* L. Pers.). In *Selected Weeds of the United States*. Agricultural Handbook No. 366, p. 86. Washington, DC: U.S. Department of Agriculture, U.S. Government Printing Office.

PROPAGATION

Johnsongrass propagates by seed and by axillary buds on rhizomes.

SPREAD

Johnsongrass spreads locally by seeds and rhizomes and to distant areas by seeds and rhizome fragments transported by various mechanical means.

DESCRIPTION

Johnsongrass is a clump-type grass with as many as 170 stems (culms) per plant. Its stems are erect and grow to a height of 3 to 10 ft (1 to 3 m), about 0.5 inch (1.25 cm) thick, pithy, with swollen nodes. The leaves are alternate, simple, smooth, 6 to 24 inches (15 to 60 cm) long and 0.5 to 1.5 inches (1.25 to 3.8 cm) wide. The leaves have about 95 stomata/sq mm^2 on the upper leaf surface and 113 on the lower surface.

In the southeastern United States, after continuous warm weather arrives in late April to early May, emergence and growth of johnsongrass plants from both seeds and rhizomes is rapid. Shoots that emerge during brief warm spells in February or March are usually killed by freezing temperatures that may occur into March.

In the fall or early winter, the aboveground vegetation of the plants is killed by early frosts and freezing temperatures. In northern areas where the ground freezes, rhizomes are also killed to the depth to which freezing temperatures penetrate the soil. The characteristic growth cycle of johnsongrass is shown in Figure 3.3.

SEEDS

Johnsongrass seeds are dark reddish-brown and nearly 0.125 inches (3.2 mm) long without the hull, with a spirally twisted awn, bent about halfway up, which is about 0.5 inch (1.25 cm) long and easily broken off. The mature seeds are shed at the point of attachment (the rachilla), and they shatter easily. The seeds fall to the soil beneath the parent plant, as there are no appendages to aid dispersal.

Reproducing freely by seed, a single plant of johnsongrass can produce as many as 80,000 seeds in one growing season. Yields of 1.1

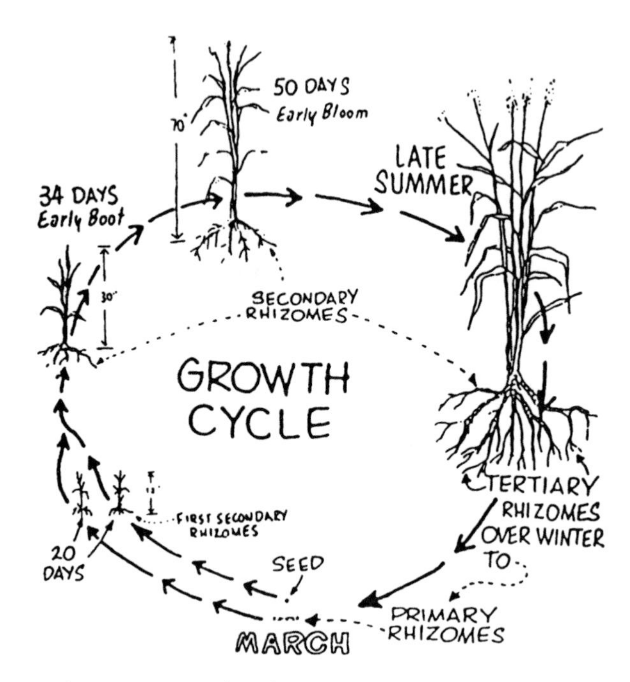

FIGURE 3.3. Johnsongrass growth cycle.

Source: Willhusen, H.W. 1961. Control Johnsongrass. Leaflet No. 250 (Rev.), Fayetteville: Agricultural Extension Service, University of Arkansas.

kg/plant have been reported in Mississippi. Seed production may vary from 482 to 1286 lb/A (540 to 1440 kg/ha).

Highest seed yields occur at photoperiods of 10.5 to 12 hours, and seedhead formation is inhibited by a photoperiod of 16 hours. Many of the seeds produced in one season are dormant, and they can remain dormant in the soil for years before germinating. The total seed production per plant is primarily dependent on the number of culms and seedheads produced, and temperature, moisture, and daylength all affect culm and seedhead production. The seedhead is a loose, open, green or purplish panicle from 6 to 24 inches (15 to 60 cm) long.

SEEDLINGS

Johnsongrass seedlings continually emerge throughout the growing season. They usually begin to develop rhizomes within 3 to 4 weeks after emergence.

PERENNATION

Rhizomes are responsible for the perennial nature of johnsongrass, surviving moderately severe winters. However, rhizomes are killed where soil temperatures fall below freezing.

RHIZOMES

Johnsongrass rhizomes are grouped as: *primary, secondary,* and *tertiary*. Primary rhizomes are those that have overwintered and from which the first shoots are produced in the spring. When the plants are 8 to 18 inches (20 to 46 cm) tall, secondary rhizomes begin to form at the base of each new plant. The primary rhizomes begin to decay 2 to 3 weeks after growth starts. Throughout the growing season, the secondary rhizomes produce the shoots that develop into the aboveground vegetation. Johnsongrass readily spreads by secondary rhizomes, and adventitious buds on even small rhizome fragments may sprout and form new plants. Tertiary rhizomes are produced at the base of mature plants in late summer. They are commonly found to a soil depth of 15 to 30 inches (0.38 to 0.76 m), sometimes as deep as 4 ft (1.2 m). The tertiary rhizomes remain dormant throughout the winter and become primary rhizomes the next year.

Johnsongrass rhizomes commonly grow to a depth of 10 to 20 inches (25 to 50 cm) in noncompacted soil. If moisture and nutrients are not limiting factors, the rhizomes will penetrate deeper in porous soils than in compacted clays. In dry clay soils, the rhizomes may penetrate to depths of 2 to 3 ft (0.6 to 0.9 m) by following cracks in the soil.

The rhizomes are stout and creeping, usually with scales at the nodes, and a freely branching, fibrous, feeder root system. The rhizomes form an extensive underground network that spreads in all directions from the original plant. Within a few years after johnsongrass becomes established, the soil is usually heavily infested with rhizomes. Under favorable conditions, rhizomes contain large reserves of stored food (carbohydrates).

Johnsongrass rhizomes are whitish to yellow, fleshy, and vary in diameter from 0.25 to 0.75 inch (0.6 to 2 cm). Each rhizome typically grows about 5 ft (1.5 m) long in tilled soil, forming nodes at intervals of 0.5 to 3 inches (1.25 to 7.5 cm) along its length, and a new plant can sprout from each node (Figure 3.4). A single johnsongrass plant can produce 130 to 295 ft (40 to 90 m) of intertwined rhizomes in a single season.

Seedling johnsongrass and plants developing from rhizomes will form a rhizome "spur" at approximately 4 weeks of age (Figure 3.5). At this time, the plants have become established, with no detectable differences between plants originating from seeds or rhizomes.

Flowering begins 6 to 7 weeks after the plants emerge from the soil, and there is little rhizome growth between the time a rhizome is initi-

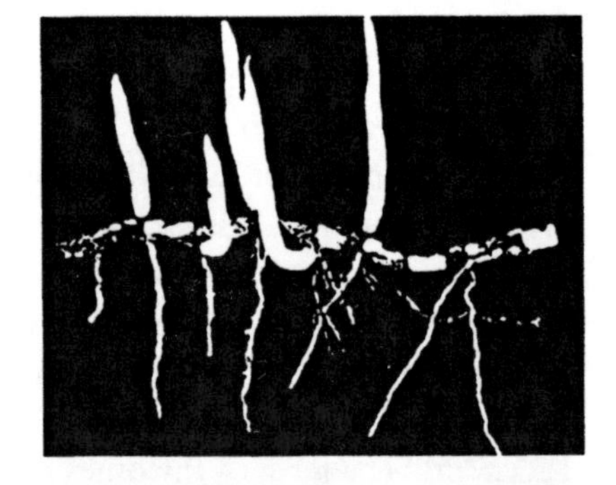

FIGURE 3.4. New johnsongrass shoots developing from axillary buds at nodes on rhizome section.

Source: Lee, O.C., and E.B. Oyer. 1954. Johnsongrass Control. Lafayette, Ind.: Agricultural Extension Service, Purdue University.

FIGURE 3.5. Johnsongrass seedling at 4 weeks of age, showing rapidly developing rhizome.

Source: Anderson, L.E. 1961. Johnsongrass in Kansas. Circular 380. Manhattan: Agricultural Experiment Station, Kansas State University.

ated and flowering occurs. Rhizome growth increases rapidly after flowering.

In Mississippi, individual plants start growth in the first week of May and produce rhizomes with an average length of 6 inches (15 cm) by the end of May. During June, July, August, and September the rhizomes grow an average of 24, 45, 100, and 43 ft (7.3, 13.7, 30.5, and 13.0 m), respectively; as much as 212 ft (65 m) of intertwined rhizomes in one growing season.

MISTAKEN IDENTITY

Johnsongrass may be mistaken for sudangrass (*Sorghum vulgare sudanese*). However, sudangrass is an annual, does not have rhizomes, and its seeds differ in color and structure from those of johnsongrass. The seeds of sudangrass are brown in color, whereas those of johnsongrass are dark reddish-brown. The seeds also differ in the structure of their pedicels (the short stalk on the seed that joins the seed to the seedhead). At maturity, the swollen tip of the pedicel detaches from the sudangrass seed by abscission, while that of johnsongrass remains attached.

PRUSSIC ACID

The young plants of johnsongrass contain the glucoside *dhurrin*, which breaks down to form *prussic acid* (hydrocyanic acid, HCN), a substance poisonous to cattle, sheep, and goats. Small plants and young tillers of johnsongrass are high in prussic acid, and its content decreases as the plants mature. The upper leaves contain more prussic acid than do the lower ones. Prussic acid content in the leaves is 3 to 25 times greater than in the stalks of plants in the "boot stage." Cattle, sheep, and goats often die when grazing on the green plants. A half-gram of prussic acid can kill a cow. Drought-stressed and second-growth plants, being small and consisting largely of leaves, are high in prussic acid, as are wilting plants following frost, freezing, cutting, or trampling. Silage, well-cured fodder, and hay may usually be fed with safety. Silage may contain toxic quantities of prussic acid, but it escapes in gaseous form when the silage is being moved.

REFERENCES

Anderson, L.E. 1969. Johnsongrass. Crop Soils 22(3): 7–9.

Anderson, L.E., A.P. Appleby, and J.W. Weseloh. 1960. Characteristics of johnsongrass rhizomes. Weeds 8: 402–406.

Anderson, W.P., J.W. Whitworth, S.S. Szabo, and W.L. Gould. 1968. Johnsongrass Control on Ditchbanks. Bulletin 527. Las Cruces: Agricultural Experiment Station, New Mexico State University, 20 pp.

Burt, G.W., and I.M. Wedderspoon. 1971. Growth of johnsongrass selections under different temperatures and dark periods. Weed Sci. 19: 419–423.

Hauser, E.W., and H.F. Hauser. 1958. Johnsongrass as a Weed. Farmers' Bulletin 1537, U.S. Department of Agriculture, Washington, DC: U.S. Government Printing Office.

Holm, L.G., D.L. Plucknett, J.V. Pancho, and J.P. Herberger. 1977. *Sorghum halepense* L. Pers. The World's Worst Weeds. Honolulu: University Press of Hawaii.

McWhorter, C.G. 1961. Morphology and development of johnsongrass from seeds and rhizomes. Weeds 9: 558–562.

McWhorter, C.G. 1971. Growth and development of johnsongrass ecotypes. Weed Sci. 19: 141–147.

McWhorter, C.G. 1971. Anatomy of johnsongrass. Weed Sci. 19: 385–393.

McWhorter, C.G. 1971. Introduction and spread of johnsongrass in the United States. Weed Sci. 19: 496–500.

McWhorter, C.G. 1972. Johnsongrass: Its history. Weeds Today 3(3): 12.

McWhorter, C.G. 1989. History, biology, and control of johnsongrass. Rev. Weed Sci. 4: 85–121.

McWhorter, C.G. 1993. A 16-yr survey on levels of johnsongrass (*Sorghum halepense*) in Arkansas, Louisiana, and Mississippi. Weed Sci. 41: 669–677.

Parker, K.F. 1972. An Illustrated Guide To Arizona Weeds. Tucson: The University of Arizona Press, pp. 72–73.

Reed, C.F., and R.O. Hughes. 1970. Johnsongrass (*Sorghum halepense*). In Selected Weeds of the United States. Washington, DC: U.S. Department of Agriculture, Superintendent of Documents, pp. 86-87.

Wax, L.M., R.S. Fawcett, and D. Isely. 1990. Johnsongrass (*Sorghum halepense*). Weeds of the North Central States. Bulletin 772. Urbana: University of Illinois at Urbana-Champaign, p. 40.

Whitson, T.D. (ed). 1991. Johnsongrass. Weeds of the West. Laramie: Western Society of Weed Science and Cooperative Extension Service of the Western States, University of Wyoming, pp. 494–495.

4

Quackgrass
(Elytrigia repens; previously, *Agropyrons repens)*

Introduction

Quackgrass, also called couchgrass, is a herbaceous, perennial grass (Figure 4.1). Quackgrass is a prohibited noxious weed in 31 states of the United States and a restricted noxious weed in 19 states. It is a plant of open areas, and it is common in agricultural fields, pastures,

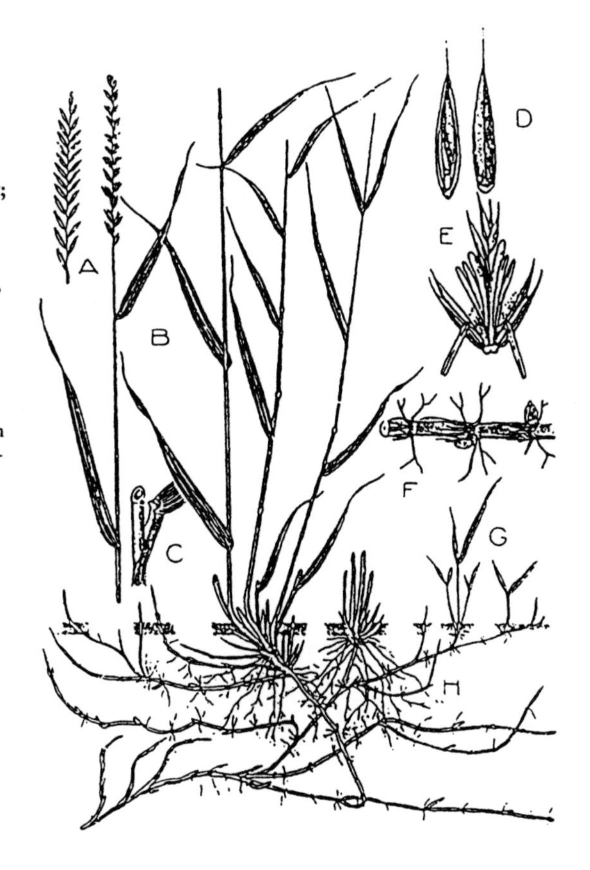

FIGURE 4.1. Quackgrass. A. Seedhead (spike); B. culm; C. ligule; D. seeds; E. floret; F. rhizome segment showing nodes, roots, and axillary buds; G. aerial shoots; H. "erect" rhizome.

Source: Anonymous. 1956. Quackgrass (*Agropyrons repens*). Leaflet 48. Bozeman: Agricultural Extension Service, Montana State College.

gardens, lawns, open waste areas, abandoned fields, and in fallow land, especially the 1st year after tillage ceases.

Quackgrass is a strong competitor with cultivated crops, reducing crop yields and quality. It seriously competes with and reduces yields of corn, soybeans, and small grains, as well as hay and pasture crops. It is a heavy consumer of key nutrients, and a stand of quackgrass can remove as much as 55%, 45%, and 68% of the total nitrogen (N), phosphorus (P), and potassium (K), respectively, from the soil. Allelopathic effects exhibited by quackgrass have been well documented. However, in some cases, the adverse allelopathic effects attributed to quackgrass may actually be partly due to successful competition of the weed with the crop for available nutrients and water. Where plentiful, quackgrass is used for pasturing and as a grass hay crop. Established stands of quackgrass hold soil against erosion on hilly land and reduce soil heaving.

DISTRIBUTION

Quackgrass is a native of Europe and was introduced into North America during colonization of New England, spreading to other areas of North America as a contaminant in hay, straw, and seed crops. Quackgrass is found in temperate climates and does not occur in the tropics.

Quackgrass is distributed throughout the United States, except for the Gulf and lower Atlantic states and most of the lower parts of Arizona and New Mexico (Figure 4.2). In 1997, plants of quackgrass were found for the first time near Soccoro and Alamogordo, New Mexico. Quackgrass thrives north of the 35th parallel, in both cultivated and noncultivated lands. It is rarely a troublesome crop weed at latitudes south of Washington, D.C., and St. Louis, Missouri.

In Canada, quackgrass is found from Newfoundland in the east to British Columbia in the west; it is especially common in southeastern Canada. In Alberta, quackgrass infests an estimated 2.2 million A (900,000 ha) of cropland, with about 30% of this land heavily infested.

PROPAGATION

Quackgrass reproduces by seeds and by axillary buds on extensively creeping, slender, straw-colored rhizomes.

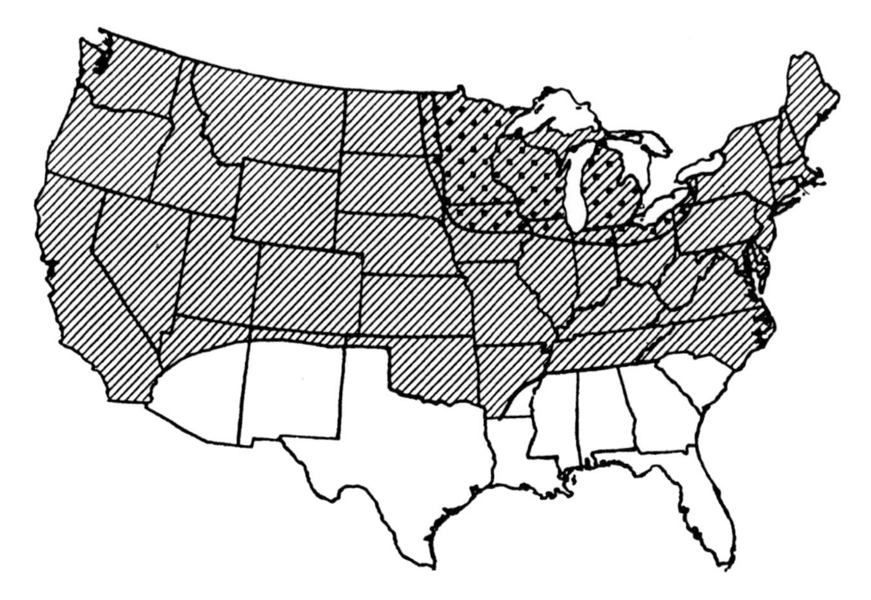

FIGURE 4.2. Distribution of quackgrass in the United States. Denser cross-hatching denotes infestations of greater economic importance.

Source: Reed, C.F., and R.O. Hughes. 1970. Quackgrass (*Agropyron repens* L. Beauv.). In *Selected Weeds of the United States*. Agricultural Handbook No. 366, p. 34. Washington, DC: United States Department of Agriculture, U.S. Government Printing Office.

SPREAD

Quackgrass spreads locally by seeds and creeping rhizomes and to distant areas by seeds.

GENETIC VARIATIONS

Quackgrass reproduction from rhizomes produces plants similar to the parent in all genetic characteristics. However, being wind pollinated and tending to be self-sterile, quackgrass is subject to great genetic variation among plants grown from seed. For example: they may be either hairy or smooth; light or dark green in color; the outer glumes and lemmas may be blunt, or have awns of varying lengths; plant height may vary from 12 to 36 inches (30 to 90 cm) and spikelet length may vary from 3 to 6 inches (7.5 to 15 cm). In some genetic lines, the spikelets remain together when harvested, while in others the spikelets break apart into individual florets making it difficult to clean them from crop seeds; some are heavy seed producers, while others are not;

some are easily killed by herbicides, while others appear resistant to such treatment.

DISTINGUISHING CHARACTERISTICS

Quackgrass is distinguished from most other grasses by the presence of matted, whitish (pale yellow or straw-colored) rhizomes, leaves with auricles, lower leaf-sheaths hairy, and seed heads resembling a slender head of wheat. The rhizomes have a tough brownish sheath at each node, giving the rhizome a scaly appearance.

Quackgrass resembles western wheatgrass (*Agropyrons smithii*), but the latter has bluish rigid leaves that tend to roll in at the edges under dry conditions, while those of quackgrass are lax, rarely bluish, and always remain flat.

DESCRIPTION

Quackgrass plants are erect, 1 to 3 ft (30 to 90 cm) high, often bending out and up at the base. Fibrous feeder roots form at the rhizome nodes, and they are relatively short compared to other grasses; they arise only at the rhizome nodes. Aerial stems are slender, smooth, with three to six nodes, and hollow at the tip; the cartilaginous bands at the upper nodes are longer than thick. The gray-green leaf blades are 0.1 to 0.3 inch (3 to 8 mm) wide and 1.6 to 12 inches (4 to 30 cm) long, with a constriction resembling an M or W about three-fourths of the way up the blade. The blades are soft, flat, with crowded fine ribs, somewhat hairy above and smooth below; at their base, there is a pair of claw-like auricles that clasp the stem. Ligules are up to 0.5 mm long. The leaf sheaths are hairy to glabrous. The seedheads are spikes 3 to 15 inches (7.5 to 15 cm) long, resembling the slender head of wheat. The plants flower in late May to September.

Quackgrass plants are most active in tillering and photosynthesis in the spring and autumn and in sexual reproduction and rhizome formation in the middle of the summer. However, the normal seasonal cycles may be altered by cultural practices.

SEEDS

In general, quackgrass produces only about 25 viable seeds per plant, but seed production varies between clones. The seeds are brownish in color and 4 to 5 mm long. The seeds possess no special appendages that aid dispersal, and they fall passively beneath the parent plant. The seeds are nondormant, they germinate in early spring, and germination is favored by fluctuating temperatures (e.g., 15°–25°C).

SEEDLINGS

In a study comparing seedling emergence from seed planted in soil at depths of 0.5, 1.0, and 1.5 inches (1.3, 2.6, and 3.8 cm), the percent seedling emergence was 73%, 11%, and 4%, respectively. Some seeds buried at greater depths in the soil germinated but died before emerging.

Quackgrass seedlings begin to produce tillers when in the four-leaf to six-leaf stage. It is usually 2 to 3 months after seed germination before rhizome buds are visible on the seedlings. Tillers and rhizomes are both initiated from axillary buds at the stem nodes of seedlings. However, tillers arise from the upper nodes (nodes 5 to 8) and rhizomes develop from the lowest nodes (nodes 1 to 4) on the stem. Each primary stem typically bears three tillers and three or four rhizomes when growing in a dense patch. Isolated plants form many secondary tillers and rhizomes; each tiller can give rise to two or three rhizomes, and each rhizome may form numerous lateral rhizomes.

Plants originating from rhizome buds begin to form new rhizomes when they are in the three-leaf to four-leaf stage and to produce tillers when in the six-leaf to eight-leaf stage.

PERENNATION

Rhizomes are responsible for the perennial nature of quackgrass, surviving moderately severe winters. However, rhizomes are killed where soil temperatures fall below freezing.

RHIZOMES

Quackgrass rhizomes are slender, about 0.125 inches (3 mm) in diameter, and at first they are white, turning pale yellow or straw-colored as they get older. The rhizomes terminate in very sharp points that easily penetrate the soil. They have also been found penetrating potato tubers and wood buried in the soil. New rhizomes start to form in the spring and may continue to do so throughout the growing season. Rhizome development is rapid under long days with adequate light, but it is reduced under short days or under low light intensity (as by shading). Greatest rhizome production occurs during July and August. During the summer months, the rhizomes grow horizontally in the soil and, in undisturbed land, nearly all the rhizomes are in the top 4- to 6-inch (10- to 15-cm) layer of soil.

Vegetative cloning is much more important than sexual reproduction in maintaining a quackgrass population. A heavy infestation of quackgrass may produce more than 35.4 tons/A (13,400 kg/ha) of rhizomes. The creeping habit of quackgrass rhizomes enable the plants to spread vegetatively and the axillary buds at the rhizome nodes provide for asexual reproduction.

Quackgrass rhizomes normally contain large amounts of food reserves, enabling them to support new growth from axillary buds even when severed from the parent plant. Rhizome segments exhibit polarity, with buds located toward the apical end forming aerial shoots, while buds located toward the base form rhizomes or remain dormant.

The longevity of individual rhizomes varies from 15 months to 2 to 3 years. In well-established quackgrass stands, the soil may become crowded with rhizomes of different ages, forming a rhizome mat. In one study, one season's growth from a single rhizome node produced 14 rhizomes having a combined total length of 458 ft (140 m), spreading over an area 11 ft (3.4 m) in diameter; 206 aerial shoots arose from this cluster of rhizomes.

Rhizomes exposed to the sun and air on the soil surface dry out quickly and die, and they also die when exposed to the dry cold air of winter at temperatures below –6°C (20°F).

Bud Dormancy

Rhizome growth is renewed annually in the spring from axillary buds located at the base of aerial shoots. It has been estimated that over 95% of the axillary buds on a rhizome remain dormant because of the strong apical dominance exerted by the terminal bud. The axillary buds are released from dormancy when the terminal bud is removed or the rhizome is severed from the parent plant.

A secondary type of dormancy, late spring dormancy, occurs during the month of June. In this type of dormancy, the lateral buds remain dormant even though environmental conditions are favorable for growth, and the rhizomes are disturbed and segmented by repeated tillage operations.

AERIAL SHOOTS

Aerial shoots are mainly produced at the end of the growing season when the rhizome tips turn upward toward the soil surface (such tips are referred to as *erects*) and undergo transformation to aerial shoots. This transformation can occur at any time during the growing season if the parent plant or rhizome is disturbed by agricultural practices or by grazing. Rhizome tips also turn upward when the plants are shaded. Aerial shoots are also formed from rhizome terminal and axillary buds whenever a rhizome is severed from the parent plant. In fall and early winter, the aerial vegetation of quackgrass is killed by early frosts and freezing temperatures.

REFERENCES

Anonymous. 1956. Quackgrass (*Agropyrons repens*). Leaflet 48, Extension Service. Bozeman: Montana State College.

Binning, L.K., R.S. Fawcett, and R.G. Harvey. 1976. Quackgrass control in vegetable crops. Proc. North Central Weed Control Conf. 31: 155.

Carder, A.C. 1963. Monuron for eradication of quackgrass. Weeds 11: 308-310.

Doll, J.D. 1993. Quackgrass Management in Field Crops. North Central Regional Extension Publication No. 219. Fargo: North Dakota State University.

Fawcett, R.S., and H.E. Davis. 1976. Effect of environment on glyphosate activity in quackgrass. Proc. North Central Weed Control Conf. 31: 159–160.

Gaines, X.M., and D.G. Swan. 1972. Quackgrass: *Agropyrons repens*. In Weeds of Eastern Washington and Adjacent Areas. Davenport, WA: Camp-Na-Bor-Lee Associations, pp. 26–27.

Harker, K.N. 1995. Short-term split application of grass-specific herbicides on quackgrass (*Elytrigia repens*) under field conditions. Weed Technol. 9: 710–715.

Harvey, R.G. 1973. Quackgrass: Friend or foe. Weeds Today 4(4): 8–9.

Harvey, R.G. 1976. Quackgrass control in forage. Proc. North Central Weed Control Conf. 31: 154–155.

Hay, J.R. 1962. Biology of quackgrass and some thoughts on its control. Down To Earth 18(1): 14–16.

Holm, L.G., D.L. Plucknett, J.V. Pancho, and J.P. Herberger. 1977. *Agropyrons repens* L. Beauv. In The World's Worst Weeds. Honolulu: University Press of Hawaii, pp. 153–168.
Johnson, B.B., and K.P. Buchholtz. 1962. The natural dormancy of vegetative buds of quackgrass. Weeds 10: 53–57.
Knake, E. 1964. Controlling Quackgrass in Illinois. Circular 892, College of Agriculture. Urbana: University of Illinois.
Laude, H.M. 1953. The nature of summer dormancy in perennial grasses. Bot. Gaz. 114: 284–292.
Ohman, J.H., and T. Kommedahl. 1964. Plant extracts, residues, and soil minerals in relation to competition of quackgrass with oats and alfalfa. Weeds 12: 222–231.
Putnam, A.R. 1976. Management of quackgrass around woody plants and turf. Proc. North Central Weed Control Conf. 31: 155–158.
Raleigh, S.M., T.R. Flanagan, and C. Veatch. 1962. Bulletin 365, Life Studies as Related to Weed Control in the Northeast, No. 4: Quackgrass. Kingston: University of Rhode Island.
Reed, C.F., and R.O. Hughes. 1970. Quackgrass (*Agropyrons repens*). In Selected Weeds of the United States. Washington, DC: U.S. Department of Agriculture, Superintendent of Documents, pp. 34–35.
Stobbe, E.H. 1976. Biology of quackgrass. Proc. North Central Weed Control Conf. 31: 151–152.
Swanton, C.J. 1985. Wirestem muhly (*Muhlenbergia frondosa* Poir Fern.). Weeds Today 16(2): 7–9.
Vengris, J. 1962. The effect of rhizome length and depth of planting on the mechanical and chemical control of quackgrass. Weeds 10: 71–74.
Wyse, D.L. 1976. Quackgrass control in field crops. Proc. North Central Weed Control Conf. 31: 152–154.

PART THREE

Perennial Grass Weeds Reproducing from Buds on Creeping Rhizomes and Stolons

5

Bermudagrass
(Cynodon dactylon)

Introduction

Bermudagrass (Figure 5.1), also called devilgrass, vinegrass, wiregrass, is a long-lived, warm-season, spreading, prostrate, herbaceous, perennial grass. It is a controversial plant. It is a restricted noxious weed in 13 states of the United States. It is considered to be one of the most serious grass weeds in the world and is ranked second among the worst weeds of the world. Its pollen is a serious source of hay fever.

Bermudagrass is a desired hay, pasture, and lawn grass in the southern-half of the United States, and it is the principal pasture grass in the southeastern United States. It is used to reduce soil erosion, bind soil on ditch and canal banks, and to stabilize soil embankments. In the early 1900s, bermudagrass was introduced into Paraquay, South America, to stabilize railroad embankments; it escaped into agricultural lands to become one of the most serious crop weeds in that country.

Bermudagrass occurs in cultivated lands, flower beds, lawns, pastures, and open places. It is a weed in arid lands, thriving in irrigated areas, and, once established, is a serious weed. It is a survivor of climatic extremes, surviving both drought and floods. It has little shade tolerance. Bermudagrass tolerates indefinite periods of drought but is unproductive under prolonged dry conditions. It is adapted to a wide range of soils, from sand to heavy clay and acid to alkaline, both arable and nonarable. It grows best on a medium to heavy soil that is moist and well drained. Its aerial vegetation is killed by frost and freezing temperatures. Bermudagrass flowers during the summer, from May to November, depending on location.

Nomenclature

There are numerous natural strains (biotypes) and improved strains (hybrids) of bermudagrass. Natural and improved strains are used as hay, pasture, and lawn grasses, and escapes of any of these strains may become a weed. The name *common bermudagrass* denotes the wild strain of bermudagrass. The general term *bermudagrass* will be used in this text to denote any form of this plant species, as all forms may become problem weeds.

FIGURE 5.1. Bermudagrass, a perennial, prostrate grass. A. Habit ×0.5; B. spike ×7.5; C. seed ×15; and D. ligules ×5.

Source: Reed, C.F., and R.O. Hughes. 1970. Bermudagrass (*Cynodon dactylon* L. Pers.). In *Selected Weeds of the United States*. Agricultural Handbook No. 366, p. 55. Washington, DC: Agricultural Research Service, United States Department of Agriculture, U.S. Government Printing Office.

DISTRIBUTION

Bermudagrass is distributed approximately throughout the southern two-thirds of the United States (Figure 5.2), but occurs as far north as central Washington and central New York, New Hampshire, and Vermont. Its range worldwide extends from latitude 45 degrees N to 45 degrees S. It was introduced into the United States prior to 1807, probably from Africa.

PROPAGATION

Bermudagrass propagates by seed and by axillary buds at the nodes of rhizomes and stolons. Some hybrids of bermudagrass are sterile and do not produce seeds; these hybrids reproduce only from axillary buds on rhizomes and stolons.

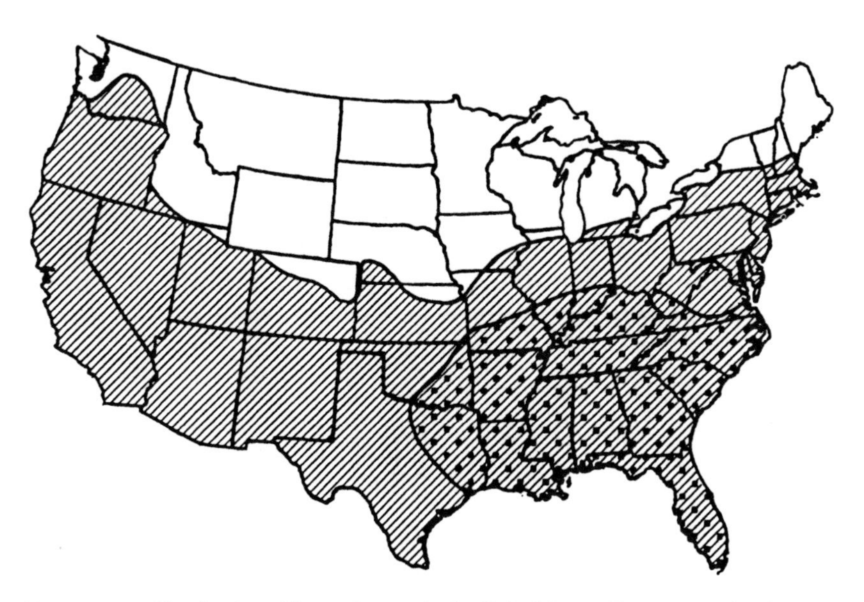

FIGURE 5.2. Distribution of bermudagrass in the United States. Denser cross-hatching denotes infestations of greater economic importance.

Source: Reed, C.F., and R.O. Hughes. 1970. In *Selected Weeds of the United States*. Agricultural Handbook No. 366, p. 54. Washington, DC: United States Department of Agriculture, U.S. Government Printing Office.

SPREAD

Bermudagrass spreads vegetatively by rhizomes and stolons, with stolons being the primary means of spread. The seeds are readily carried by wind and irrigation water, and they may be a contaminant in commercial crop seeds, hay, and in livestock feed and bedding. Bermudagrass is also spread artificially by rhizome-fragments being dragged from place to place by cultivating machinery or carried as contaminants in mud on machinery and animal hooves.

SEEDS

Bermudagrass is a very sparse seed producer, except in the southwestern United States where, in Arizona and southern California, it is grown as a seed crop. Seedset in the southeastern United States is less than 1% (essentially zero). In Arizona and southern California, seedset is as much as 95%, apparently favored by high temperatures and low humidity. The seed is tiny (1.5 mm long), oval, awnless, and orange-red, reddish-brown, or straw-colored, numbering about 2 million/lb (4.4 million/kg). The seed is free within the lemma and palea.

The seeds can remain dormant in the soil, and they maintain their viability well. The seeds pass through the digestive tract of cattle and sheep, retain their viability, and may even show improved germination. Bermudagrass seed will germinate when daily temperatures are above 65°F (18°C). Optimum vegetative growth occurs when mean temperatures are above 75°F (24°C). Aerial vegetative growth is killed by temperatures below 28°F (2°C).

DESCRIPTION

Bermudagrass is a long-lived, prostrate, perennial grass that spreads by stolons and a vast system of hard, sharp-pointed rhizomes. Fibrous roots arise from nodes on rhizomes and stolons. Stems also arise from the nodes on rhizomes and stolons. The stems are weak, variable in internode length, and curve upward from the spreading base. Only the flowering stems or stem tips are erect, 6 to 18 inches (15 to 45 cm) tall, terminating in a seedhead 6 to 18 inches (15 to 45 cm) tall. Decumbent stems spread laterally over the soil surface, rooting freely at lower nodes in contact with moist soil.

Leaf blades are 1 to 4 inches (2.5 to 10 cm) long, ⅛ inch (3 mm) wide, gray-green, and glabrous except for a fringe of long hairs along the edge just above the collar. The ligule is a ring of conspicuous white hairs. The leaf sheaths are up to ⅝ inch (15 mm) long, shorter than the stem internodes, and glabrous except for tufts of hairs on either side of the collar.

DISTINGUISHING CHARACTERISTICS

The distinguishing characteristics of bermudagrass are the sharp-pointed rhizomes, stolons rooting readily at nodes in contact with moist soil, gray-green foliage, conspicuous white-hairy ligule at the base of the leaf blade, fringe of hairs on the keel of the lemma, two rows of seeds tightly appressed along one side of the rachis, with the seeds slightly overlapping, and the one whorl of three to seven, usually five, fingerlike branches (rachises) arising from the same point at the tip of the flowering stem.

MISTAKEN IDENTITY

Bermudagrass is sometimes confused with large crabgrass (*Digitaria sanguinalis*) because of a superficial resemblance between inflorescences and because both have prostrate spreading stems that root at the nodes when in contact with moist soil. They may be correctly identified by noting the arrangement of their seeds along their rachises. Also, bermudagrass is a perennial, with rhizomes and stolons; large crabgrass is an annual without rhizomes and stolons.

The stalkless, slightly overlapping seeds of bermudagrass are tightly appressed in two rows along one side of the somewhat flattened or triangular rachis. In contrast, the seeds of large crabgrass are arranged in two rows along opposite sides of the rachis, even though they originate on the same side; seeds in one row are without pedicels, and those in the other row have pedicels about 1.25 to 1.75 mm long.

In addition, the inflorescence of bermudagrass has three to seven, usually five, slender fingerlike branches (rachises), 0.8 to 2.4 inches (2 to 6 cm) long, clustered in a single whorl at the tip of the flowering stem. The inflorescence of large crabgrass has three to 11 slender fingerlike branches (rachises), 2 to 6 inches (5 to 15 cm) long; these may arise at the same point at the stem tip, forming a whorl, but with several rachises usually located on the stem a short distance below the whorl at the tip.

PERENNATION

Rhizomes are the basis for the perennial habit of bermudagrass, surviving moderately severe winters.

RHIZOMES

The rhizomes of bermudagrass are hard, scaly, sharp-pointed and form a dense heavy sod. New plants grow from the axillary buds located at the rhizome nodes. A single bud on a rhizome or rhizome fragment may develop into a shoot. The basal node on this shoot has lateral buds which give rise to tillers or rhizomes according to an established pattern. Rhizomes are found at soil depths of a few inches to 3 ft or more (a few centimeters to a meter or more); many rhizomes have buds which can sprout from below the plow layer, about 8 inches (20 cm). Where there is a cold winter season, reserve carbohydrates accumulate in the roots and rhizomes through autumn and into midwinter. These reserves are then used to support new growth in the spring.

STOLONS

The stolons are the principal means by which bermudagrass spreads. They are flat, hairless, extensively creeping, and bear at each node the dead, bladeless leaf sheaths that make the "dog's teeth," which give the plant its Latin name, *Cynodon dactylon.* The stolons vary in length from a few inches up to 5 ft (a few cm to 1.5 m). Fibrous roots and new shoots grow from stolon nodes in contact with moist soil.

HYDROCYANIC ACID

Bermudagrass vegetation may contain hydrocyanic acid (prussic acid) under some conditions. The percentage of toxicant is high following frost or pronounced drought with high temperatures. When ingested, hydrocyanic acid can be poisonous to cattle, sheep, and goats.

REFERENCES

Anonymous. 1945. Bermudagrass (*Cynodon dactylon*). Grass: The Yearbook of Agriculture. Washington, DC: U.S. Department of Agriculture, U.S. Government Printing Office, pp. 663–664.

Elmore, C.D. 1985. Bermudagrass (*Cynodon dactylon*). Weed Identification Guide. Champaign, IL: Southern Weed Science Society, CYNDA, p. 2.

Fischer, B.B., A.H. Lange, and B. Crampton. 1985. Bermudagrass (*Cynodon dactylon* L. Pers.). Sheet WI-72. In Growers Weed Identification Handbook. Richmond: Cooperative Extension Service, University of California.

Holm, L.G., D.L. Plucknett, J.V. Pancho, and J.P. Herberger. 1977. The World's Worst Weeds. Honolulu: The University Press of Hawaii, pp. 25–31.

Jordan, T.N. 1977. Today's weed: Bermudagrass. Weeds Today 8(3): 10.

Kneebone, W.R. 1968. Bermudagrass found world-wide. Progressive Farmer Arizona 20(3): 21, 23.

Kneebone, W.R. 1968. Bermuda covers the globe. Progressive Farmer Arizona 20(4): 16–17.

Manning, E. 1975. Callie bermudagrass. Progressive Farmer. June, p. 18.

Parker, K.F. 1972. An Illustrated Guide to Arizona Weeds. Tucson: The University of Arizona Press, pp. 34–35.

Reed, C.F., and R.O. Hughes. 1970. Bermudagrass (*Cynodon dactylon*). In Selected Weeds of the United States. Bulletin No. 366. Washington, DC: U.S. Department of Agriculture, Superintendent of Documents, pp. 54–55.

Robbins, W.W., M.K. Bellue, and W.S. Ball. 1951. Weeds of California. Sacramento, CA: State Office of Publications, pp. 68–69.

Part Four

Grasslike Perennial Weeds

A. Creeping Perennials Reproducing from Buds on Tubers

6

PURPLE NUTSEDGE
(Cyperus rotundus)

YELLOW NUTSEDGE
(Cyperus esculentus)

INTRODUCTION

Purple nutsedge and yellow nutsedge (Figure 6.1) are herbaceous, grasslike perennials. They are two of the most troublesome perennial weeds of cultivated lands wherever they grow. Purple nutsedge has the infamous distinction of being named "the world's worst weed." In the United States, the two nutsedge species rank fifth in seriousness among all weed species and second only to quackgrass among perennial weed problems. Purple nutsedge is designated a prohibited noxious weed in the seed laws of 14 states, and yellow nutsedge is so designated in the seed laws of 10 states.

Both nutsedge species are found in cultivated fields, lawns, gardens, ditch banks, roadsides, and waste areas. They can grow upward through asphalt pavement, penetrate plastic swimming pool liners, and grow through potato tubers and storage roots of root crops. They reduce crop yields and quality, and interfere with crop harvesting.

Purple nutsedge and yellow nutsedge grow with reckless abandon in almost every soil type, humidity, and acidity/alkalinity (pH), and they survive the hottest temperatures known in agriculture. Neither species tolerates shade, but their dormant tubers remain viable, ready to sprout and reinfest the area when conditions are again favorable for growth.

Due to the cold winters, purple nutsedge does not grow in the midwestern United States (the Corn Belt). In contrast, yellow nutsedge is the most troublesome perennial weed in agronomic crops in the Corn Belt, severely infesting an estimated 6.4 million acres (2.6 million ha) of corn and 4.2 million acres (1.7 million ha) of soybeans. In the northeastern United States, it infests more than 3.5 million acres (1.4 million ha) of corn.

FIGURE 6.1. Plants of purple nutsedge (A) and yellow nutsedge (B), showing inflorescence, flower stalk, leaves, and characteristically shaped tubers.

Source: Robbins, W.W., M.K. Bellue, and W.S. Ball. 1951.Weeds of California, p. 116. Sacramento: California State Department of Agriculture, State Office of Documents and Publications.

Purple nutsedge does not grow in Canada due to the extreme cold temperatures. However, yellow nutsedge is found in most of southeastern Canada, and it is a persistent weed in many fields of vegetables, corn, and oats.

DISTRIBUTION

Purple nutsedge and yellow nutsedge are found throughout the tropic, subtropic, and temperate zones. In North America, purple nutsedge is found almost as far north as the 35th parallel (roughly on a line east to west through Washington, D.C., St. Louis, and San Francisco). Its range at higher latitudes is apparently limited by cold temperatures.

Yellow nutsedge is a weed on all continents. In the Western Hemisphere, it is found from southern Canada to northern Argentina. Yellow nutsedge is found in every state of the United States and in the Canadian provinces of Ontario, Quebec, New Brunswick, and Nova Scotia. It is a serious weed in the 12 northeastern states of the United States. Infestations of yellow nutsedge continue to increase nationwide. Over the past 30 years, the author has seen yellow nutsedge become the dominant nutsedge species in cotton, onion, lettuce, and chile pepper crops in southern New Mexico, far exceeding that of purple nutsedge, which had been the dominant nutsedge species. The distribution of yellow nutsedge in the United States is shown in Figure 6.2.

PROPAGATION

Both purple nutsedge and yellow nutsedge propagate primarily by axillary buds on tubers and basal bulbs. Although purple nutsedge produces seeds, few are viable, and they contribute little to its propagation. Yellow nutsedge produces viable seed, but the small, nonvigorous seedlings tend not to survive in cultivated fields.

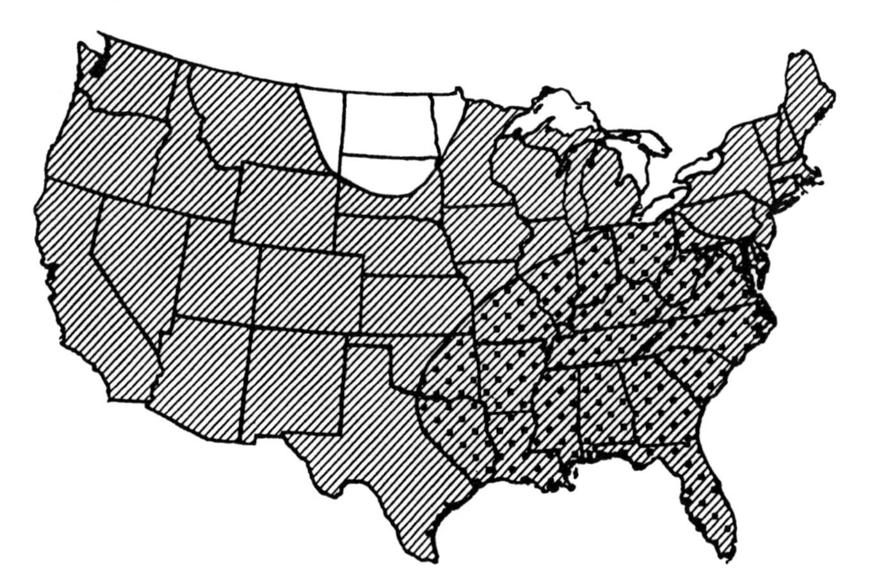

FIGURE 6.2. Distribution of yellow nutsedge in the United States. Denser cross-hatching denotes infestations of greater economic importance.

Source: Reed, C.F., and R.O. Hughes. 1970. Yellow nutsedge (*Cyperus esculentus* L.). In *Selected Weeds of the United States*. Agricultural Handbook No. 366, p. 97. Washington, DC: United States Department of Agriculture, U.S. Government Printing Office.

SPREAD

Purple nutsedge and yellow nutsedge spread naturally by creeping rhizomes. They spread artificially by their tubers and basal bulbs being dragged from place to place by cultivating tools, carried in soil attached to farm equipment, and transported as stowaways in the soil of ornamental and vegetable transplants, among other means of transportation. Yellow nutsedge spreads also by seed dispersal.

PERENNATION

Tubers are responsible for the perennial nature of both purple nutsedge and yellow nutsedge, surviving moderate to severe winters.

DESCRIPTION

Purple nutsedge and yellow nutsedge plants are erect, grasslike, perennial sedges. They have unjointed, triangular, solid stems (flower stalks), and their leaves are three-ranked with closed sheaths and without ligules. Plants of purple nutsedge and yellow nutsedge have single flower stalks that bear an umbellike inflorescence. Both purple nutsedge and yellow nutsedge possess the C_4-dicarboxylic-acid-CO_2-fixation pathway. The stomates are nearly all on the lower leaf surface; those on the upper surface are near the leaf margins.

The leaves of purple nutsedge are 0.1 to 0.3 inch (0.25 to 0.75 cm) wide, varying in length from 2 to 8 inches (5 to 20 cm) and are seldom longer than the flower stalk.

The leaves of yellow nutsedge are 0.25 to 0.5 inch (0.6 to 1.3 cm) wide, with slightly roughened edges. They are as long or longer than the 6 to 12 inches (15 to 30 cm) high flower stalk.

Purple nutsedge plants consist of an underground network of chains of tubers joined by rhizomes and, aboveground, rosettes of leaves and umbel-bearing scapes (flower stalks). At the junction of a rhizome and leaves, a tuberous enlargement develops, which is termed the *basal bulb*.

In general, yellow nutsedge plants are similar to purple nutsedge, with the primary exceptions being that its inflorescence is of a different color, each rhizome terminates in a tuber, and chains of rhizome-connected tubers are not formed. A simplified diagram of the life cycle of yellow nutsedge in the Corn Belt is given in Figure 6.3.

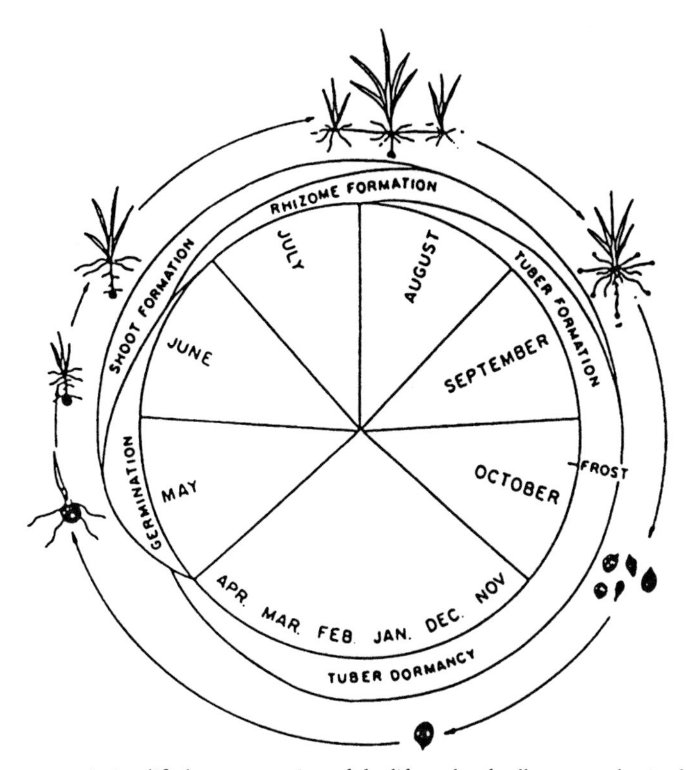

FIGURE 6.3. A simplified representation of the life cycle of yellow nutsedge in the midwestern United States (the Corn Belt). The width of the band enclosing the five major processes indicates the intensity of the process. Tuber dormancy is depicted only during the nongrowing season, but it occurs throughout the year.

Source: Stoller, E.W. 1981. Yellow Nutsedge: A Menace in the Corn Belt. Technical Bulletin No. 1642. Washington, DC: U.S. Department of Agriculture, U.S. Government Printing Office.

DISTINGUISHING CHARACTERISTICS

Purple nutsedge and yellow nutsedge can be visually distinguished, one from the other, by the color of their umbellike inflorescence and leaves, and by the color, shape, and arrangement of their tubers. The inflorescence of purple nutsedge is reddish to purple-brown, while that of yellow nutsedge is yellow to yellowish-brown. Purple nutsedge and yellow nutsedge take their names from the color of their respective inflorescence.

The leaves of purple nutsedge are glossy, dark-green, while those of yellow nutsedge are shiny, yellowish-green. The leaves of both nutsedge species have a prominent mid-vein.

Morphologically, the shape of the leaf tip of yellow nutsedge is distinctive and often offers a means for distinguishing this species from purple nutsedge. In yellow nutsedge, the leaf has a "shoulder" 0.4 to 0.8 inch (1 to 2 cm) from the tip, which tapers to an attenuated, needlelike point. In comparison, purple nutsedge leaves taper rather abruptly to a blunt tip.

Purple nutsedge tubers tend to be elliptical in shape, varying in size from about 0.75 to 1.0 inch (2.0 to 2.5 cm) in length and 0.5 to 0.75 inch (1.3 to 2.0 cm) in diameter. There are eight to 10 buds per tuber, one bud per node. The tubers are white when newly formed, turning black as they mature. They have a musty, tannin-like, bitter taste and a lingering, pungent after-flavor. Most importantly, the tubers are interconnected, one to another, by rhizomes to form connected chains of tubers, with as many as 10 tubers per chain.

Yellow nutsedge tubers tend to be spherical, ranging in size from about 0.1 to 0.5 inch (3 to 12 mm) in diameter. There are five to seven buds per tuber, with one bud per node. The tubers are white when newly formed, turning brown as they mature. The tubers have a pleasant, nutlike flavor similar to almond. Most importantly, the tubers terminate their respective rhizome; thus, chains of tubers are not formed. The longevity of individual yellow nutsedge tubers in soil is relatively short, with over 80% of the tubers decaying in less than 3 years.

In the United States, most purple nutsedge and yellow nutsedge tubers are found in the top 6-inch (15-cm) soil layer and rarely below 12 inches (30 cm).

It is recognized that there are ecotypes of both purple nutsedge and yellow nutsedge, but a systematic classification of the ecotypes has not been made.

Seed Production

In one 2-year study, purple nutsedge produced an average of 186 flowers per inflorescence the 1st year and 1354 flowers per inflorescence the second year of the study. Only 43 seeds were found and none germinated.

In contrast, yellow nutsedge produces an abundance of viable seed, as much as 1500 seed per inflorescence with better than 50% viability. In one study, it was estimated that yellow nutsedge could produce 11 to 125 million seeds/A (27 to 309 million/ha). In Canada, yellow nutsedge produces few, if any, viable seed, apparently due to the colder climate.

APICAL DOMINANCE

In the case of purple nutsedge, the apical bud of a single tuber always sprouts first. In a rhizome chain of tubers, the tuber at the morphological apex (tip) of the chain prevents sprouting by other tubers in the chain). This dominance is not as strong as that exerted by the apical bud within a tuber itself. The separation of a tuber from a chain of tubers removes it from the apical dominance of the apical tuber in the chain; an important consideration when tillage operations severe the tuber chains and drag and relocate tubers within the field. With yellow nutsedge, tubers are not connected in chains, but bud germination in the tuber is governed by apical dominance as with tubers of purple nutsedge.

DORMANCY

Tubers of purple nutsedge and yellow nutsedge formed during the 1st year are usually dormant, but tubers which have overwintered in the field show a high percentage (95%) of sprouting in the spring.

TUBER PRODUCTION

Tubers of purple nutsedge planted at 12-inch (30-cm) intervals produced the equivalent of 3.1 million plants and 4.4 million tubers and basal bulbs per acre (7.7 million plants and 11 million tubers and basal bulbs/ha). Tubers planted at 36-inch (90-cm) intervals produced 2.3 million plants and 2.8 million tubers and basal bulbs per acre (5.7 million plants and 7 million tubers and basal bulbs/ha). New tubers were produced 21 days after parent tubers were planted.

By the end of the 2nd year, the purple nutsedge tuber population increased by 66% where spacings were 12 inches (30 cm) and by 84% with the 36-inch (90-cm) spacings. Purple nutsedge may produce as much as 36,000 lb/A (40,000 kg/ha) of underground plant material. In a field study in Georgia, it was estimated that purple nutsedge produced over 10 million basal bulbs and dormant tubers per acre (25 million/ha) during one growing season from tubers spaced 36 inches (90 cm) apart.

One tuber of yellow nutsedge planted in a field produced 36 plants and 332 tubers in 16 weeks, and, within 1 year, 1900 plants and almost 7000 tubers were produced, forming a patch 7 ft (6 m) in diameter; equivalent to a tuber yield of 8.3 tons (fresh weight) per acre (18.6 thousand kg/ha). In Georgia, a yellow nutsedge colony formed from one tuber field-planted produced 622 tubers in 17 weeks.

BASAL BULBS

The basal bulb is considered the principal site of vegetative activity and propagation of both purple nutsedge and yellow nutsedge. The basal bulb is a compressed, stemlike structure consisting of a series of closely appressed leaves and sheathing. It arises from a tuberous base and is connected to the tuber by a rhizome. Basal bulbs are formed at the tips of rhizomes that have turned upward toward the soil surface, The basal bulbs develop when the tips of the vertical, or near vertical, rhizomes are 2 to 5 inches (5 to 10 cm) below the soil surface, but some may develop deeper. These bulbs contain meristems for fibrous roots, rhizomes, leaves, and flowers. The basal bulb in both purple nutsedge and yellow nutsedge is the base from which the leafy shoot and the subterranean rhizome system develop.

Rhizomes grow from the basal bulbs and, at very short or long distances from the parent bulb, their tips turn upward and terminate in secondary basal bulbs. Thus, by repetition, it is possible for the primary basal bulb to give rise to a series of secondary, tertiary, and higher-order basal bulbs and vegetative shoots, each in turn producing more rhizomes and tubers.

RHIZOMES

Rhizomes are the means by which purple nutsedge and yellow nutsedge spread vegetatively, and it is through the rhizomes that photosynthates produced in the leaves move to the tubers and basal bulbs. In cross section, the rhizome is seen to have an epidermis, cortex, endodermis, pericycle, xylem, and phloem. The central vascular bundles (xylem and phloem) are continuous through the tuber and basal bulb network. The rhizome has an apical meristem, but there are apparently no buds at the nodes along the rhizome. Thus, little or no propagation is possible from rhizome fragments. The tubers of both purple nutsedge and yellow nutsedge form as swellings at the tips of rhizomes. Rhizome growth terminates with the formation of a tuber or basal bulb. Rhizome growth is indeterminate, some growing to 24 inches (60 cm) in length, while the growth of others is so short as to appear that the basal bulb has arisen directly from the tuber.

FIBROUS ROOTS

Fibrous roots originate in the endodermal tissues of tubers, basal bulbs, and rhizomes, penetrating the cortex and epidermis, with no arrangement with respect to nodes or other physical features. They constitute only a small portion of the total plant biomass. Rhizomes, basal bulbs, and tubers dominate the subterranean part of the plant.

REFERENCES

Lewis, W.M. 1972. Today's weed: Nutsedges. Weeds Today 3(3): 19.

Loustalot, A.J., T.J. Muzik, and H.J. Cruzado. 1954. Studies on nutgrass (*Cyperus rotundus* L.) and its control. Bulletin 52. Mayaguez: Federal Experiment Station in Puerto Rico.

Mulligan, G.A., and B.E. Junkins. 1976. The biology of Canadian weeds, 17: *Cyperus esculentus*. Can. J. Plant Sci. 56: 339–350.

Nandihalli, U.B., and L.E. Bendixen. 1988. Toxicity and site of uptake of soil-applied imazaquin in yellow and purple nutsedge (*Cyperus esculentus* and *C. Rotundus*). Weed Sci. 36: 411–416.

Reed, C.F., and R.O. Hughes. 1970. Yellow nutsedge (*Cyperus esculentus*) and purple nutsedge (*C. Rotundus*). In Selected Weeds of the United States. Washington, DC: U.S. Department of Agriculture, Superintendent of Documents, pp. 96–99.

Stoller, E. W. 1975. Growth, development, and physiology of yellow nutsedge. Pro. North Central Weed Control Conf. 30: 124–125.

Stoller, E.W. 1981. Yellow nutsedge: A menace in the Corn Belt. Technical Bulletin No. 1642. Washington, DC: U.S. Department of Agriculture, Superintendent of Documents.

Stoller, E.W. and L.M. Wax. 1973. Yellow nutsedge shoot emergence and tuber longevity. Weed Sci. 21: 76–81.

Sweet, R.D. 1975. Control of nutsedge in horticultural crops. Proc. North Central Weed Control Conf. 30: 129–130.

Taylorson, R.B. 1967. Seasonal variation in sprouting and available carbohydrates in yellow nutsedge tubers. Weed Sci. 15: 22–24.

Thullen, R.J., and P.E. Keeley. 1979. Seed production and germination in *Cyperus esculentus* and *C. rotundus*. Weed Sci. 27: 502–505.

Tumbleson, M.E., and T. Kommedahl. 1962. Factors affecting dormancy in tubers of *Cyperus esculentus*. Bot. Gaz. 123: 186–190.

Tumbleson, M.E., and T. Kommedahl. 1963. Reproductive potential of *Cyperus esculentus* by tubers. Weeds 9: 646–653.

Tweedy, J.A., A.J. Turgeon, and D.W. Black. 1975. Control of yellow nutsedge in turf. Proc. North Central Weed Control Conf. 30: 131–132.

Wax, L.M. 1975. Control of yellow nutsedge in field crops. Proc. North Central Weed Control Conf. 30: 125–128.

Wills, G.D. 1975. Taxonomy, morphology, anatomy, and composition of yellow nutsedge. Proc. North Central Weed Control Conf. 30: 121–124.

Wills, G.D. 1977. Nutsedge deals misery to cotton growers. Weeds Today 8(2): 16–17.

Wills, G.D., and G.A. Briscoe. 1970. Anatomy of purple nutsedge. Weed Sci. 18: 631–635.

Wills, G.D., R.E. Hoagland, and R.N. Paul. 1980. Anatomy of yellow nutsedge (*Cyperus esculentus*). Weed Sci. 28: 432–437.

B. Noncreeping Perennials Reproducing from Bulbs

7

Wild Garlic
(Allium vineale)

Wild Onion
(Allium canadense)

Introduction

Wild garlic (Figure 7.1) is a ubiquitous, bulbous, herbaceous, perennial monocot. It was probably introduced to the United States from Europe, possibly France, by early settlers in the early 1700s for food flavoring and/or inadvertently as a contaminant in wheat seed. Wild onion (Figure 7.2) is a native of North America, and it is distributed more widely in the United States than is wild garlic.

Wild garlic and wild onion are cool-season weeds, with a growing season extending from fall to spring. They are drought-hardy, cold-hardy, tolerant to wet soils, and grow best in heavy soils. They are often found growing on the same sites. They are troublesome weeds in fields of small grain pastures, hay crops, and in lawns, gardens, along roadsides, and in noncrop areas. They are especially troublesome in the midwestern and eastern United States where soft red winter wheat is grown. Wild garlic and wild onion cause identical crop damage, imparting a "garlicky" odor and flavor to dairy products and meat.

Distribution

The distribution of wild garlic is shown in Figure 7.3. Wild garlic is found as far south as Alabama, Georgia, and Mississippi, as far north as Massachusetts, New York, Ohio, and Michigan, as far west as Colorado, eastern Kansas, and Oklahoma, with localized infestations reported in central Wyoming and along the Pacific Coast in northern California, Oregon, and Washington.

Wild onion is distributed more widely in the north central and northeastern United States than is wild garlic. It is found as far north

FIGURE 7.1. Wild garlic: A. Habit ×0.5; B. flower cluster ×0.5; C. aerial bulblets ×2.5; D. flower ×3.5; E. hard-shell bulbs ×0.5; soft-shell bulbs ×0.5.

Source: Reed, C.F., and R.O. Hughes. 1970. Wild garlic (*Allium vineale* L.). In *Selected Weeds of the United States*. Agricultural Handbook No. 366, p. 107. Washington, DC: United States Department of Agriculture, U.S. Government Printing Office.

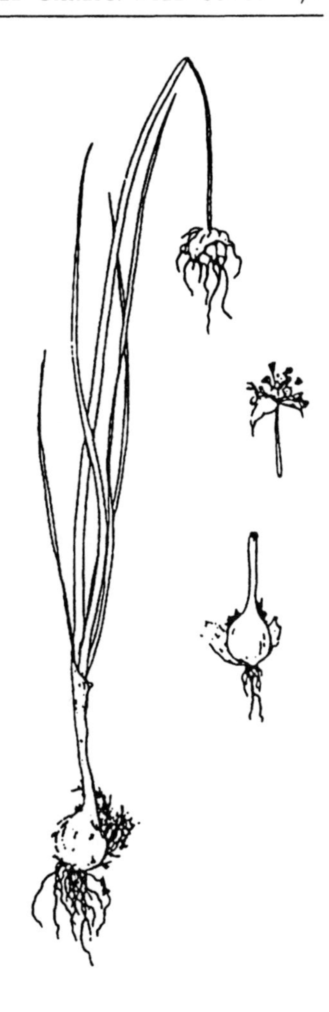

FIGURE 7.2. Wild onion, showing entire plant on left, flower cluster in upper right, and old bulb in lower right.

Source: Weeds of the North Central States. North Central Regional Research Publication No. 281. 1981. Urbana: University of Illinois at Urbana-Champaign.

as Canada, Minnesota, and Wisconsin, and as far south as Texas and Florida. Wild onion is not found in states located to the west of a general line extending from western Minnesota south to east Texas.

PROPAGATION

Reproduction of wild garlic in the United States is almost entirely asexual, by means of various bulb types. The reproductive cycles of wild

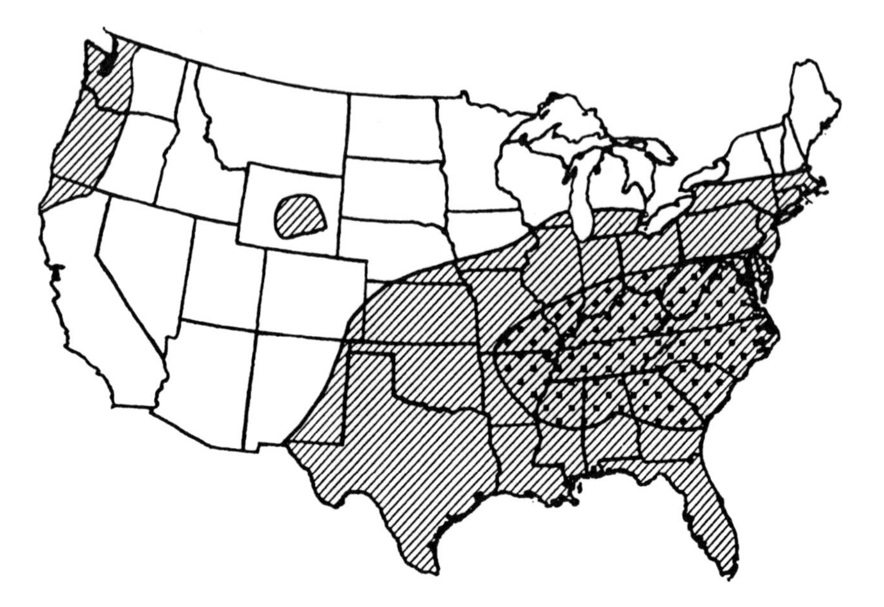

FIGURE 7.3. Distribution of wild garlic in the United States. Denser cross-hatching denotes infestations of greater economic importance.

Source: Reed, C.F., and R.O. Hughes. 1970. Wild garlic (*Allium vineale* L.). In *Selected Weeds of the United States*. Agricultural Handbook No. 366, p. 106. Washington, DC: United States Department of Agriculture, U.S. Government Printing Office.

garlic showing the relationship between bulb types, seed, and plant types are shown in Figure 7.4. Bulb types give rise to either scapigerous or nonscapigerous plants (see Plant Types below). Seeds give rise to nonscapigerous plants the 1st year and to either type of plant after the 1st year. Seed production is insignificant in most wild garlic habitats in the United States, except near the southern limits of its range where viable seed production is abundant (e.g., Virginia and Delaware). When seeds are produced, they are usually viable.

SPREAD

Wild garlic usually spreads by bulbs rather than seeds, and most usually by aerial bulblets.

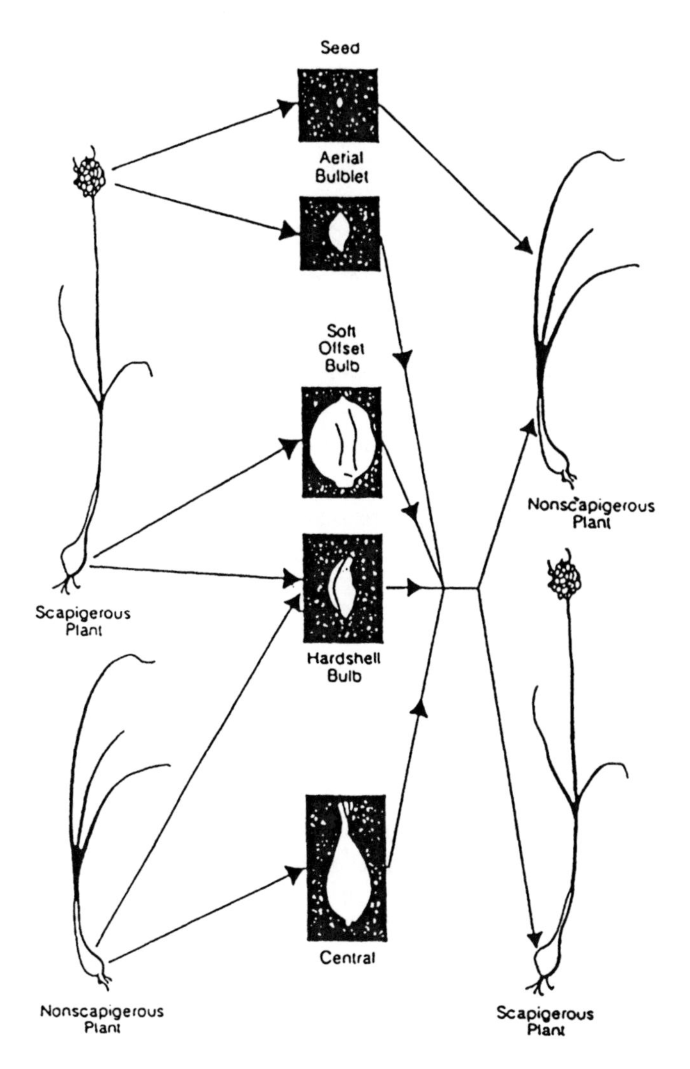

FIGURE 7.4. Reproductive cycles of wild garlic showing relationship between bulb types, seed, and plant types. Bulb types give rise to either scapigerous or nonscapigerous plants. Seed give rise to nonscapigerous plants the 1st year and to either type later.

Source: Peters, E.J. 1976. Wild Garlic: Life Cycle and Control. Agricultural Information Bulletin No. 390. Washington, DC: U.S. Department of Agriculture, U.S. Government Printing Office.

Description

Wild garlic plants look much like cultivated onions. Wild garlic is a bulbous herb, the outer leaves of the bulb formed from the foliar sheathing leaves. The *leaves* are two-ranked, mostly 4 to 8 inches (10 to 20 cm) tall. The leaf blades are circular, hollow in cross section, striped, with the younger blades easily flattened and slenderly tapering. The flower stems are smooth, waxy, and 1 to 3 ft (30 to 90 cm) tall. The flowers are greenish-white, small, on short stems above aerial bulblets, wholly or partially replaced with bulblets. The *spathe* (a large, dry, thin, sheathing bract partly enclosing the inflorescence), usually consists of one bract rather than a pair. It is short, beaked, and the edges are united above. The inflorescence projecting through the base of the deciduous spathe is nearly heart-shaped and 1 to 2 inches (2 to 5 cm) in diameter. The seeds are black, flat on one side, and about 0.125 inch (3 mm) long. Seeds produced in the spring germinate the following fall.

Distinguishing Characteristics

Wild garlic, wild onion, and star-of-Bethlehem appear similar to the casual observer, and they are often misidentified. However, there are identifiable differences.

Star-of-Bethlehem (*Ornithogalum umbellatum*) is a herbaceous, perennial monocot, resembling wild garlic and wild onion and often found growing on the same sites. It is often planted as an ornamental and then spreads to lawns, gardens, and adjoining areas.

Wild garlic can be distinguished from wild onion and star-of-Bethlehem by its striate (minutely grooved), hollow leaves, attached at the lower half of the plant. Wild onion and star-of-Bethlehem have flat, solid leaves attached at the base of the plant. The leaves of star-of-Bethlehem have a white stripe down their centers. The stems of wild garlic are 1 to 3 ft (30 to 90 cm) tall; those of wild onion 1 to 2 ft (30 to 60 cm) tall; those of star-of-Bethlehem are seldom more than 8 in (20 cm) high. The flowers of wild garlic are small and greenish-white, those of wild onion are pink, and those of star-of-Bethlehem are showy white.

Wild garlic has underground hard-shell bulbs, which are absent from wild onion and star-of-Bethlehem. Wild garlic has offset bulblets, while wild onion has none. The old bulb coat of wild garlic is thin and membranous, while that of wild onion is fiber-matted. Star-of-Bethlehem does not have the "garlicky" odor of wild garlic and wild onion.

Wild garlic and wild onion begin growth in mid-August or early September and mature in late May or early June. Star-of-Bethlehem begins growth soon after the ground thaws in early spring; small, showy, white flowers appear, and then the plants mature and disappear before warm weather.

UNIQUE CHARACTERISTICS

Wild garlic and wild onion are unusual in that they pose a serious threat to *crop quality,* with little effect on *crop yield.* Aerial bulblets produced by wild garlic are about the same size as wheat kernels and are difficult to separate when harvested with the grain. When milled with the grain, the bulblets are crushed and impart a garlicky flavor to the flour and, due to their high moisture content, gum up the milling equipment, causing the machinery to be shut down and cleaned. Garlic-infested grain is usually sold as feed grain.

Wild garlic foliage also imparts a "garlicky" odor and flavor to milk, butter, cheese, and even meat when ingested by grazing dairy cows. The objectionable garlicky odor and flavor lies with allyl sulfide [$(CH_2{=}CHCH_2)_2S$], a chemical that occurs in all parts of the wild garlic plant and is the chief constituent of the oil of garlic.

PLANT TYPES

Wild garlic populations consist of two plant types: *scapigerous* and *nonscapigerous.* Scapigerous plants have flower stalks (scapes) that produce aerial bulblets. The flowering stalks sometimes produce flowers that have a greenish or purplish perianth. One scape may produce as many as 300 aerial bulblets. If seeds are produced, they appear among the aerial bulblets and on the scape in the spring and germinate in the fall. The smaller, nonscapigerous plants do not produce a scape at the end of the growing season, and they do not produce aerial bulblets.

BULB TYPES

Four types of wild garlic bulbs are found at the end of the growing season in late spring. They are *aerial bulblets, hard-shell bulbs, central bulbs,* and *soft offset bulbs.* All four types of bulbs have a short after-ripening period following senescence of the parent plant. During this after-ripening period from early June to late July, the root and leaf primordia elongate slowly. At the conclusion of the after-ripening period, these primordia elongate rapidly. Each type of bulb is capable of producing either scapigerous or nonscapigerous plants. A scapigerous plant produces seeds and aerial bulblets aboveground plus six hard-shell bulbs and one soft offset bulb belowground. A nonscapigerous plant produces one central bulb and sometimes one or two hard-shell bulbs at maturity.

The aerial bulblets are the most numerous of the four types, and they consist of a fleshy, cone-shaped scale with a growing point at its base. The fleshy scale is a bladeless storage leaf.

The hard-shell bulbs are the second most numerous, and they are larger than the aerial bulblets. The hard-shell bulbs are formed underground in the axils of the outer leaves of both scapigerous and nonscapigerous plants. Similarly, they have a single bladeless storage leaf with a growing point at its base. The storage leaf is surrounded by a bladeless leaf that forms a hard protective shell about the bulb. The hard-shell bulb is the only bulb type produced by both scapigerous and nonscapigerous plants. With hard-shell bulbs, little or no elongation occurs during the after-ripening period until late August when about 20% to 35% of the hard-shell bulbs break dormancy and sprout. The remaining hard-shell bulbs break dormancy slowly and some may remain dormant up to 6 years.

The central bulb is formed underground by nonscapigerous plants, and it is conspicuous in the spring at the end of the growing season. It is formed around the main axis of the plant, circular in cross section, and varies in size from the smallness of an aerial bulblet up to that of a soft offset bulb. The structure of a central bulb is similar to other bulb types, but it sometimes does not have an outer protective scale, being surrounded only by the withered bases of the foliage leaves. When the outer scale is present, it is prolonged into a sharp terminal point.

The soft offset bulbs are formed underground in the axil of the innermost leaf of scapigerous plants. They are usually the largest of the four bulb types and similar in structure to the other types.

BULB SPROUTING

Aerial bulblets that complete development in the spring (May and June) sprout in the fall of the same year. Sprouting of hard-shell bulbs starts in mid-August or early September and nearly ceases by October or November. Central and soft offset bulbs start sprouting in early fall. However, many shoots of sprouted bulbs do not emerge from the soil until the following spring (March and April).

REFERENCES

Anderson, L.E., and E.J. Peters. 1982. Today's weed: Wild garlic. Weeds Today 13(4): 7–8.

Cardina, J. 1990. Wild garlic (*Allium vineale*). Sheet 2ALLCA. In Weed Identification Guide. Champaign, IL: Southern Weed Science Society.

Ferguson, G.P., G.E. Coats, G.B. Wilson, and D.R. Shaw. 1992. Postemergence control of wild garlic (*Allium vineale*) in turfgrass. Weed Technol. 6: 144–148.

Hardcastle, W.S. 1976. Chemical control of wild garlic *Allium species*. Agron. J. 68: 144–145.

Knake, E.L., and C.R. Howell. 1976. Controlling wild garlic, Down To Earth 31(4): 21–22.

Peters, E.J. 1975. Wild garlic: A tough pest. Weeds Today 6(4): 13–15.

Peters, E.J. 1976. Wild garlic: Life Cycle and Control. Agricultural Information Bulletin 390. Washington, DC: U.S. Department of Agriculture, Superintendent of Documents.

Peters, E.J., and R.A. McKelvey. 1982. Herbicides and dates of application for control and eradication of wild garlic (*Allium vineale*). Weed Sci. 30: 557–560.

Reed, C.F., and R.O. Hughes. 1970. Wild garlic (*Allium vineale*). In Selected Weeds of the United States. Agricultural Handbook No. 366, Washington, DC: U.S. Department of Agriculture, Superintendent of Documents.

Retzinger, J. 1990. Wild onion (*Allium canadense*). Sheet 2ALLCA. In Weed Identification Guide. Champaign, IL: Southern Weed Science Society.

Spenser, E.R. 1974. Wild garlic (*Allium vineale*). In All About Weeds. New York: Dover Publications, pp. 74–77.

Spritzke, J.F., and E.J. Peters. 1970. Dormancy and sprouting cycles of wild garlic. Weed Sci. 18: 112–114.

Spritzke, J.F., and E.J. Peters. 1972. Anatomy of wild garlic bulbs during and subsequent to after-ripening. Weed Sci. 20: 233–237.

Wax, L.M., R.S. Fawcett, and D. Isely. 1990. Wild garlic (*Allium vineale*) and wild onion (*Allium canadense*). In Weeds of the North Central States. Regional Research Publication No. 281, Bulletin 772. Urbana: University of Illinois at Urbana-Champaign, p. 46.

Whitson, T.D. (ed). 1991. Wild onion and wild garlic. In Weeds of the West. Laramie: Western Weed Science Society, Bulletin Room, University of Wyoming, pp. 372–375.

Part Five

Simple Perennial Broadleaved Weeds Reproducing from Taproots and/or Root Crowns (Caudexes)

8

Dandelion
(Taraxacum officinale)

Introduction

Dandelion (Figure 8.1) is a low, tufted, simple, herbaceous, perennial broadleaf weed. It is a controversial plant; an attractive flowering plant to some and a weed to others. Most people enjoy the dandelion's colorful yellow flowers. The plant's young leaves are fancied by many as salad greens, they can also be cooked and eaten as one would spinach. In the manufacture of "dandelion wine," the flowers are used to impart a bright yellow color to the finished product. Dandelion is nearly a ubiquitous weed. Its propensity to grow in lawns and its tenacity as a weed make dandelion plants particularly odious to the suburbanite who likes to manicure his lawn on weekends.

Distribution

Dandelion is widespread throughout most of North America. Its distribution in the United States is shown in Figure 8.2. Dandelion was introduced to North America by the early European settlers as a garden plant. It grows at sea level to an elevation of about 12,000 ft (3660 m). Dandelion grows in disturbed places, such as vacant lots, fallow fields, roadsides, and in relatively undisturbed areas such as lawns, meadows, mountain slopes, and over great expanses of the Arctic area.

Propagation

Dandelion reproduces by seeds and by new growth from the root crown. New plants can arise from buds on the fleshy taproot following damage and segmentation of this organ, as from cultivation.

Spread

Wind dispersal is the principal means by which dandelion spreads. Its seeds can be carried long distances on air currents. Dandelion seed is often spread as an impurity in seeds of Kentucky bluegrass and forage grasses, and they have often been deliberately carried from place to place for cultivation. Dandelion may also be spread by physically moving segments of its fleshy taproot from one moist area to another.

FIGURE 8.1. Dandelion. A. Habit ×0.5; B. flower ×3; C. achenes ×7.5; D. achenes with pappus ×1.

Source: Reed, C.F., and R.O. Hughes. 1970. Dandelion (*Taraxacum officinale* Weber). In *Selected Weeds of the United States*. Agriculture Handbook No. 366, p. 439. Washington, DC: U.S. Department of Agriculture, U.S. Government Printing Office.

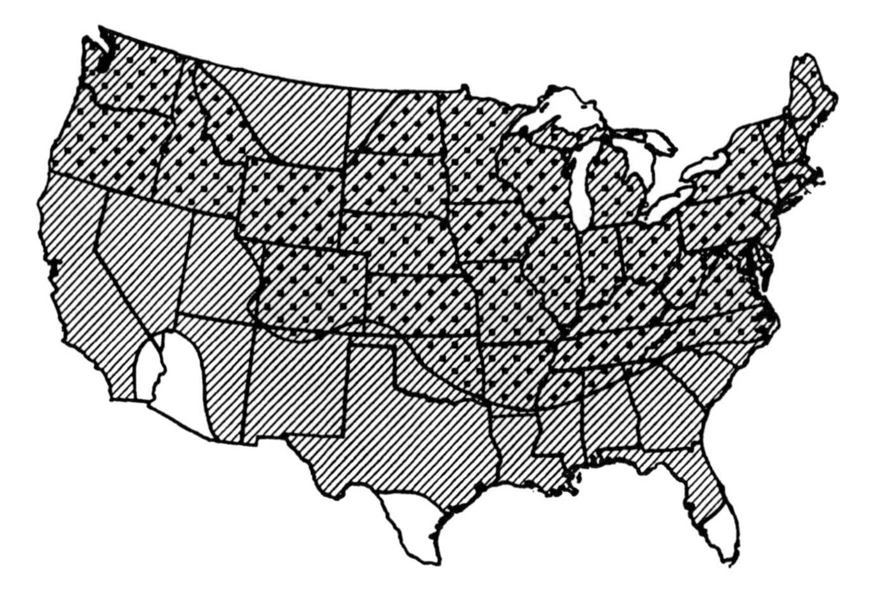

FIGURE 8.2. Distribution of dandelion in the United States. Denser cross-hatching denotes infestations of greater economic importance.

Source: Reed, C.F., and R.O. Hughes. 1970. Dandelion (*Taraxacum officinale* Weber). In *Selected Weeds of the United States*. Agriculture Handbook No. 366, p. 438. Washington, DC: U.S. Department of Agriculture, U.S. Government Printing Office.

PERENNATION

The fleshy taproot, with the characteristic of a very short, vertical rhizome, is the basis for the perennial nature of dandelion. The plants store reserve foods in their deep, stout taproots. They survive mild winters as a rosette of leaves and severe winters by their dormant taproots.

DISTINGUISHING CHARACTERISTICS

Dandelion plants are characterized by having a long, thickened, branched taproot 12 inches (30 cm) or more in length and up to 1.2 inches (3 cm) in diameter; no true stems (the stem tissue does not elongate, forming the root crown); white milky sap in all plant parts; and leaves that are clustered at the base of the plant, forming a rosette lying prostrate on the ground or ascending to upright. The white milky sap readily exudes from cut or broken plant parts.

If the dandelion taproot is injured, cut, or broken, an undifferentiated callous tissue is formed to seal the wound. Following formation of callous tissue, one or more buds are formed on this tissue from which new leaves or an entire new plant develop. One to five new plants may arise from this callous tissue. A segment of a dandelion taproot, placed in a damp atmosphere, will develop buds at the apical pole of the segment and roots at the basal pole. If the segment is cut into smaller pieces, each piece forms buds at the apical pole and roots at the basal pole. The taproot of chicory (*Chicorium intybus*) responds in a similar manner.

The clustered leaves of dandelion arise from the root crown located at or slightly below the soil surface. The leaves vary greatly in size and lobing. They are 2 to 12 inches (5 to 30 cm) long and usually divided into few or several indistinct pairs of lobes, which are pointed or blunt at the tips. Often the lobe at the tip of the leaf is much larger and triangular in shape.

Dandelion normally flowers from March to first frost or, in warmer southern areas, throughout the year. The flower heads are round, solitary, 1 to 2 inches (2.5 to 5 cm) across, and composed of golden-yellow petallike flowers or rays supported at the end of a long, naked, hollow flower stalk (called a peduncle or scape), which varies greatly in length, 3 to 24 inches (7.5 to 60 cm) or more. The strap-shaped ray flowers are five-notched at the tip, and there are 100 to 300 ray flowers per head. Bracts (phyllaries) are green to brownish and surrounding the flower head in two rows, with the outer row hanging down and one-third to one-half as long as the inner, erect row.

The dandelion is a short-day plant, blooming when there is less than 12 hours of light. However, a few occasional plants can be seen blooming in the long days of midsummer.

The actual process of flowering is very rapid, and the first sign of this process is the formation of a bud in the middle of the rosette. Bud development takes approximately a week. Then, the peduncle (scape) starts forming between the base of the bud and the tip of the shoot. As this stalk elongates, the bud is thrust upward, opens, and blooms. The entire process of flower stalk formation usually takes about 48 hours. The flower remains open one day, on the average, and then closes. Following blooming, the flower stalk and flower usually flatten to the ground. After a few days, the flower stalk straightens again, and the leaves surrounding the closed flower open revealing the round white ball of seeds, each with a "parachute" of hairs.

SEED

Dandelions produce seed asexually, and the embryo develops without fertilization. Most dandelion pollen grains are abortive, sterile, and unable to form a pollen tube. The early stages of ovule formation are normal and so is the first one-half of mitosis (i.e., splitting the chromosomes in half). However, the mitotic process is interrupted at this point and a "restitution nucleus" is formed. This nucleus has the normal 24 chromosomes of a dandelion plant, and it continues to develop into an embryo and a seed without recourse to fertilization. The result of asexual reproduction is the production of offspring that are genetically identical (barring mutations).

The seeds (achenes) are tannish, sometimes reddish-brown, about 0.1 to 0.2 inch (3 to 4.5 mm) long, five to eight ribs on each side, and minutely toothed with tiny curved spines along the upper margins. The seed ends in a long slender beak two or four times as long as the body of the achene and is attached to a tuft of persistent, fine, silky, whitish, parachutelike hairs (pappus) 3 to 4 mm long.

SEEDLINGS

Dandelion seed germinates, forming a primary root and two cotyledons. The first true leaf forms just above the cotyledons, at the end of a very short internode (about 1 mm). All subsequent leaves are also separated, one from the other, by extremely short internodes. The leaves form a spiral, and do not exactly overlap (only every sixth leaf will overlap). Although the stem does not elongate, it does increase in width up to about 1.2 inches (3 cm) wide. The leaves are spirally arranged on this flat crown, with the largest on the outside. Under undisturbed conditions, this leaf arrangement leads to the formation of a globose rosette, likened to a flattened upside-down bowl. Where the plants are subjected to mowing or trampling, as in lawns, the leaves form the more familiar flat rosette.

Soon after germination, the primary root becomes dominant and, as more leaves are formed, it increases in size, keeping pace with the shoot. Eventually, a large fleshy taproot is formed that serves as a storage organ, similar to a carrot. The taproot contains a white milky sap that is also found in all other dandelion plant parts.

REFERENCES

Anonymous. 1990. Common dandelion (*Taraxacum officinale*). In Weeds of the North Central States, Bulletin 772. Urbana: College of Agriculture, University of Illinois at Urbana-Champaign, p. 237.

Harrington, H.D. 1967. Common dandelion (*Taraxacum officinale*). In Edible Native Plants of the Rocky Mountains. Albuquerque: The University Press of New Mexico, pp. 99-103.

Korsmo, E. 1954. *Taraxacum officinale* Web. In Anatomy of Weeds. Forlag, Norway: Grondahl and Sons, pp. 396-403.

Parker, K.F. 1972. Dandelion (*Taraxacum officinale*). In An Illustrated Guide To Arizona Weeds. Tucson: The University Press of Arizona, pp. 320-321.

Reed, C.F., and R.O. Hughes. 1970. Dandelion (*Taraxacum officinale*), In Selected Weeds of the United States. Agriculture Handbook No. 366. Washington, DC: U.S. Department of Agriculture, U.S. Government Printing Office, pp. 438-439.

Solbrig, O.T. 1971. The population biology of dandelion. Am. Sci. 59: 686-694.

9

Curly Dock
(Rumex crispus)

Introduction

Curly dock (Figure 9.1) is a simple perennial broadleaf weed with a strong, deep, branched, taproot. Stems are stout, erect, and often reddish. It was introduced to North America from Europe and is naturalized in the United States and Canada. Curly dock is a serious weed in cereal crops, pastures, and clover and alfalfa fields grown for seed. It is a vigorous weed on moist soils and a common weed in cultivated crops, hayfields, meadows, gardens, lawns, irrigation ditches, roadsides, and waste areas. In the United States, curly dock is listed as a restricted noxious weed in the seed laws of 26 states.

There are several weeds of the genus *Rumex* known as "docks."

FIGURE 9.1. Curly dock (*Rumex crispus*). Basal portion of plant with leaves and large taproot. Also upper portion of fruiting branch; a. fruiting calyx and b. achene (seed).

Source: Illustrations by Lucretia B. Hamilton from *An Illustrated Guide to Arizona Weeds* by Kittie Parker. Copyright ©1972 The Arizona Board of Regents. Reprinted by permission of the Univerity of Arizona Press.

Curly dock and broadleaf dock (*Rumex obtusifolius*) are the most competitive of these. Curly dock and broadleaf dock are often confused for one another.

DISTRIBUTION

Curly dock is found throughout the United States (Figure 9.2) and southern Canada. It is a troublesome world weed in both arable lands and pastures. It has been described as one of the 12 most successful, noncultivated, colonizing weed species in the world.

PROPAGATION

Curly dock reproduces by seed and by shoots developing from buds on the root crown. Regeneration of new shoots occurs only on the upper 4-cm portion of the taproot, or from buds on root segments taken from this part; the remainder of the taproot cannot produce new shoots.

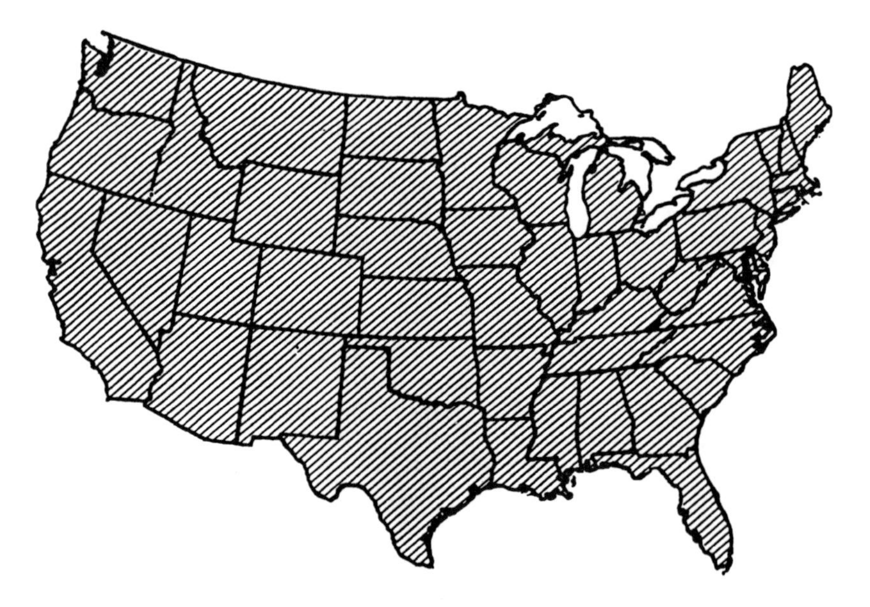

FIGURE 9.2. Curly dock distribution in the United States.

Source: Reed, C.F., and R.O. Hughes. 1970. In *Selected Weeds of the United States*. Agriculture Handbook No. 366, p. 130. Washington, DC: U.S. Department of Agriculture, U.S. Government Printing Office.

Spread

Curly dock spreads primarily by seed dispersal. The fruits are lightweight and may be carried long distances by wind. They are also spread by moving water and by adhering with mud onto moving objects. However, the seeds tend to fall near the plant and germinate there.

Curly dock seed is a common contaminant in crop seeds, such as cereals, clover, and alfalfa, and planting of such contaminated seed is the principal means by which curly dock is introduced into croplands.

Curly dock can also spread as the result of tillage practices, with regeneration from taproot segments as small as 1 inch (2.5 cm) in length.

Perennation

Curly dock overwinters as a rosette or by underground root crowns. Curly dock seedlings that emerge in autumn form an overwintering rosette from which new shoots arise in the spring.

Plant Description

Curly dock is a simple, herbaceous, perennial broadleaf weed. During the 1st year of growth, the plant forms a dense rosette of leaves on a stout taproot. The taproot is large, thick, fleshy, yellow, and somewhat branched. It may extend 5 ft (1.5 m) into the soil and branch laterally as much as 3 ft (1 m). The stems are singular or in groups from buds on the root crown. They are erect, often reddish, slender, grooved, glabrous, 1 to 4 ft (0.3 to 1.2 m) tall, simple or with a few branches at the top. The stems die back each fall, and new ones arise from the root crown each spring.

Curly dock stems have a thick, fleshy, underground portion above the root crown which is about 1.2 to 1.6 inch (3 to 4 cm) long and 2 inches (5 cm) or more wide. The underground stem portion results from root contraction, which pulls the crown down into the soil by the time the fifth true leaf is being formed.

The leaves are mostly basal, lanceolate, dark-green, alternate, up to 4 inches (10 cm) wide, and 4 to 12 inches (10 to 30 cm) long with wavy margins; upper leaves are alternate and usually less wavy than the basal leaves. The larger leaves are rounded to nearly heart-shaped at the base. The petioles are short, 1 to 2 inches (2.5 to 5 cm) long, with a papery sheath at their base surrounding the stem (as in all dock species).

The long cylindrical inflorescence terminates the stem. It has many ascending or erect branches and a few to many small, linear leaves intermingled among the branches. The small, greenish flowers have no petals, no nectar, and are wind pollinated. They are arranged in whorls or dense clusters 8 to 16 inches (20 to 40 cm) long, with ascending racemes on pedicels 5 to 10 mm long. Flowering generally occurs in April or May, but it can occur from April through October. The inflorescence becomes reddish-brown at maturity. Occasionally, plants may form two inflorescences, one in the spring and another in the fall.

The calyx consists of six greenish sepals. The three inner sepals are enlarged, heart-shaped, veiny bracts, about 5 mm long, with smooth margins. Each bract has an enlarged, corky, oblong, thickened, wort-like center called a *tubercle* or *callous grain*. The three inner sepals (in fruits, called *valves*) and accompanying tubercles form the fruit, and they enclose a single seed (achene). The seed is triangular in shape and sharp edged, about 2 mm long, and glossy brown at maturity.

Curly dock is conspicuous in early spring when it makes lush vegetative growth in the rosette stage leading to flower production. Later in the spring and summer, curly dock is easily identifiable by its showy reddish-brown inflorescence.

SEEDS

The number of seeds produced annually varies widely, from 100 to more than 60,000 per plant. Seedset occurs from June through September. Much of the seed produced in a given year is viable and capable of germination shortly after maturity. Seed germination can occur in any month of the year under proper conditions, but in general, a flush of seedlings emerges in the spring or fall.

The seeds require exposure to light, alternating temperatures, or both, for germination. Near-red light (6100 to 7000 A) stimulates germination, and far-red light (greater than 7000 A) inhibits germination. Within limits, the reaction is reversible. Darkness inhibits seed germination. However, scarification of the seed coat with acid or sandpaper greatly increases percent germination in darkness.

Curly dock seed buried in soil for 1 year germinated 80%, and germination was 76% for seeds buried for 10 years. A few curly dock seed, buried in soil, may remain viable for more than 80 years. Curly dock seedlings are poor competitors with other weeds, but once the plants become established, they are difficult to control or eradicate because of their deep taproots. Under favorable conditions, curly dock can flower and produce seeds in its seedling year.

Curly dock seeds are killed when fed to chickens, but they pass through the digestive tracts of cattle and other birds without being harmed. The seeds are short-lived when stored in silage.

MISTAKEN IDENTITY

Curly dock is sometimes confused with broadleaf dock (*Rumex obtusifolius*). Curly dock is distinguished from broadleaf dock by its long leaves narrowed at the base with prominently crinkly and wavy leaf margins, short petioles, bracts with smooth (entire) margins surrounding the seed (achene), and each of the three bracts has a tubercle.

Broadleaf dock differs from curly dock in having leaves that are broad, with a heart-shaped base, long petioles, and the undersurface of petioles and leaf veins covered with short, blunt, whitish hairs, and the three inner sepals bear three to five teeth along each margin. Only one of the three sepals has a tubercle.

The fruits of curly dock and broadleaf dock are illustrated in Figure 9.3; note the smooth margins of the sepals of curly dock and the toothed margins of those of broadleaf dock.

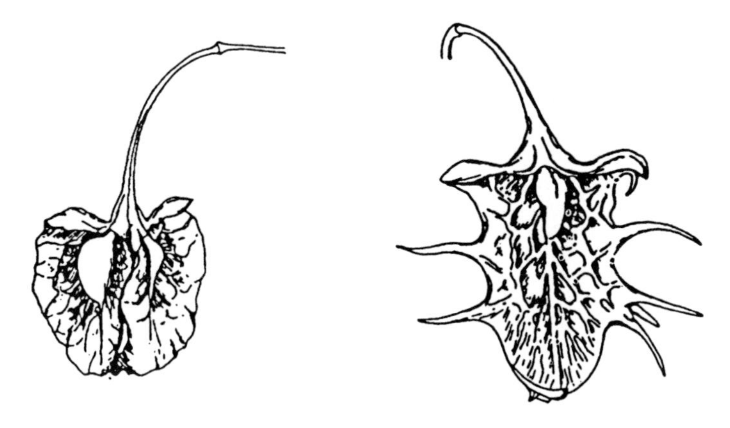

FIGURE 9.3. Fruits of curly dock (*left*) and broadleaf dock (*right*). Note the smooth margins of the bracts (sepals) curly dock and the toothed margins of those of broadleaf dock.

Adapted from Weeds of California, p. 140. 1951. Sacramento: State Department of California.

Regeneration of new shoots of broadleaf dock occurs only from buds on the upper 7.5-cm portion of the taproot, or from buds on root segments taken from this part; the remainder of the taproot cannot produce new shoots.

REFERENCES

Elmore, C.D. 1985. Curly dock (*Rumex crispus* L.). In Weed Identification Guide. Champaign, IL: Southern Weed Science Society.

Fisher, B.B., A.H. Lange, and J. McCaskill. 1985. Curly dock (*Rumex crispus* L.). In Growers Weed Identification Handbook. Davis: Division of Agricultural Sciences, University of California, p. WI-51.

Holm, L.G., D.L. Plucknett, J.V. Pancho, and J.P. Herberger. 1977. *Rumex crispus* L. and *Rumex obtusifolius* L. In The World's Worst Weeds. Honolulu: The University Press of Hawaii, pp. 401–408.

Korsmo, E. 1954. *Rumex crispus* L. In Anatomy of Weeds. Forlag, Norway: Grondahl and Sons, pp. 94–97.

Monaco, T.J. 1972. Today's weed: Curly dock (*Rumex crispus*). Weeds Today 3(2): 19.

Murphy, T.M. (ed). 1993. Curly dock (*Rumex crispus* L.). In Weeds of Southern Turf Grasses. Athens: College of Agriculture, University of Georgia, p. 166.

Parker, K.F. 1972. Curly dock (*Rumex crispus* L.). In An Illustrated Guide to Arizona Weeds. Tucson: The University of Arizona Press, pp. 90–91.

Reed, C.F., and R.O. Hughes. 1970. Curly dock (*Rumex crispus* L.). In Selected Weeds of the United States. Washington, DC: U.S. Department of Agriculture, Superintendent of Documents, pp. 130–131.

Robbins, W.W., M.K. Bellue, and W.S. Ball. 1951. Curly dock (*Rumex crispus* L.). In Weeds of California. Sacramento: State Department of California, p. 140.

Wax, L.M., R.S. Fawcett, and D. Isely. 1990 (reprint). Curly dock (*Rumex crispus)*. In Weeds of the North Central States, Bulletin 772. Urbana: College of Agriculture, University of Illinois at Urbana-Champaign, p. 55.

Whitson, T.D. (ed). 1991. Curly dock (*Rumex crispus* L.). In Weeds of the West. Laramie: Western Society of Weed Science, University of Wyoming, pp. 514–515.

10

Broadleaf Plantain
(Plantago major)

Buckhorn Plantain
(Plantago lanceolata)

Blackseed Plantain
(Plantago rugelii)

Introduction

Broadleaf plantain (Figure 10.1A), buckhorn plantain (Figure 10.1B), and blackseed plantain (Figure 10.1C) are stemless (with stemlike flower stalks), herbaceous, perennial broadleaf weeds. These plantain species are weeds of cultivated fields, gardens, lawns, meadows, pastures, roadsides, and waste areas. They are especially troublesome weeds in lawns. Buckhorn plantain prefers dry sites over moist areas, and it has been described as highly drought resistant. In Colorado, buckhorn plantain is a more pernicious weed than broadleaf plantain. It is a troublesome weed in alfalfa, clover, and grass fields, and a serious contaminant in the seeds of these crops. Buckhorn plantain is listed as a restricted noxious weed in the seed laws of 34 states in the United States, and categorized as a secondary noxious weed (Class 3) in the Canada Seeds Act and Regulations.

Distribution

Broadleaf plantain and buckhorn plantain were introduced to North America by early Eurasian settlers. Broadleaf plantain was known to be present in New England in 1672 and eastern Canada in 1821. Broadleaf plantain populations in North America may consist of both native populations and those naturalized from Eurasia. Blackseed plantain is native to eastern North America.

Broadleaf plantain and buckhorn plantain are common weeds

throughout the United States (Figures 10.2A and 10.2B). Blackseed plantain is a common weed in the eastern half of the United States, extending from Georgia, west to central Texas, Nebraska, and South Dakota, and north into southeastern Canada (Figure 10.2C).

FIGURE 10.1. A. Broadleaf plantain: a. seed; b. seed pod. B. buckhorn plantain: a. seed, note characteristic spike with prominent stamens jutting out from it. C. blackseed plantain: 1, whole plant: 2. mature seed pod; 3. seeds.

Source: Illustrations by Lucretia B. Hamilton from *An Illustrated Guide to Arizona Weeds* by Kittie Parker. Copyright ©1972 The Arizona Board of Regents. Reprinted by permission of the Univerity of Arizona Press; Weeds of the North Central States, Bulletin 772, Urbana: University of Illinois at Urbana-Champaign.

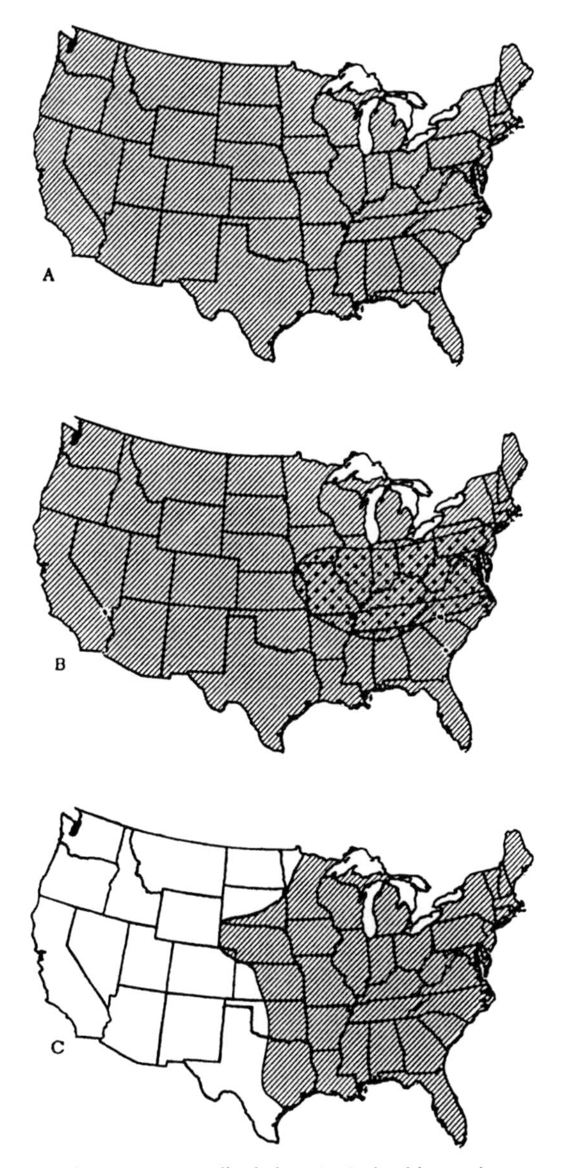

FIGURE 10.2. Distribution: A. Broadleaf plantain; B. buckhorn plantain (denser cross-hatching denotes infestations of greater economic importance); and C. blackseed plantain in the United States.

Source: Reed, C.F., and R.O. Hughes. 1970. Plantains: Buckhorn (*Plantago lanceolata* L.), broadleaf (*Plantago major* L.), and blackseed (*Plantago rugelii* Decne.). In *Selected Weeds of the United States*. Agricultural Handbook No. 366, pp. 346, 347, and 348. Washington, DC: U.S. Department of Agriculture, U.S. Government Printing Office.

Broadleaf plantain is found throughout Canada. Buckhorn plantain (known as narrow-leaved plantain in Canada) occurs across the southern portion of Canada, except in the Prairie Provinces. It is particularly abundant in southern British Columbia, Ontario, Quebec, and the coastlines of Nova Scotia and Prince Edward Island. It was first reported in Nova Scotia in 1829. Blackseed plantain (known as rugelii plantain in Canada) is a common weed in southeastern Canada from Ontario eastward to Nova Scotia. In Canada, these plantain species can be described as annual, biennial, or perennial herbs.

Broadleaf plantain and buckhorn plantain are common weeds in most of the agricultural areas of the world. They are temperate-zone plants with extreme ranges to the north and south. They have also become pestiferous weeds in almost all tropical crops. They are seldom reported as a principal weed in a particular crop due to their lack of size and vigor. On a world basis, buckhorn plantain is said to be one of the 12 most successful, noncultivated, colonizing weed species.

Common Characteristics

The following characteristics are common to broadleaf plantain, buckhorn plantain, and blackseed plantain:

Roots

In the 1st year, the seedling plant has a spindle-shaped (rodlike with tapered ends), sparingly branched root that later disappears in the 2nd year and is replaced by a short rhizome with fibrous adventitious roots (Korsmo, 1954). The root system is composed of many slender, fibrous, lateral roots. The mature root system may exceed 30 inches (80 cm) in depth and 30 inches (80 cm) in diameter.

Leaves

The leaves are alternate in rosette, simple, basal, erect, or spreading.

Flowers

The flowers are wind pollinated, numerous, inconspicuous, and densely arranged along a spike that terminates a slender, leafless flower stalk. The flowers are self-compatible. The plants are long-day plants, and they do not flower unless the photoperiod is greater than 13 hours. In general, they flower from May to October, earlier if the daylength requirement is met, but mainly in June, July, and August.

Reproduction

Reproduction is by seed, rarely by vegetative means.

Seeds

The seeds are small, 1 to 2.5 mm long, depending on species. Light is necessary for seed germination, although a small proportion of seeds may germinate in total darkness. The seeds are coated with mucilage when wetted. This adhesive mucilage on the seed coat causes it to swell, increasing the original seed size by about 40%. The gelatinous mucilage (an acid polysaccharide) is readily separated from the seed, and it has been used as a stiffener of muslin and other fabrics in France.

Dehiscence

The seed pod (capsule) splits open when mature. This is due to contraction of the fruit wall as the capsule ripens. However, there is no forceful dehiscence, and the seeds must be shaken from the spike, as by wind, rain, or animals; if not, they remain attached to the spike until it rots and falls to the ground.

VARIETIES AND HYBRIDS

Many varieties or ecotypes of these weed species have been reported; the result of varying or unfavorable growing conditions. All attempts to produce artificial hybrids have failed.

SPREAD

Each of the three species spreads by seed dispersal, primarily as contaminants in small crop seeds, such as alfalfa, clovers, and grasses. When wet, the seeds adhere to clothing, hair, fur, shoes, and other objects by sticky mucilage. They are also carried in soil adhering to vehicles and farm equipment.

PERENNATION

Each of these plantain species has a root crown (caudex) located just below the soil surface which serves as a food storage organ and from which adventitious roots and the leaves arise. The root crown is responsible for the perennial nature of these plantain species. Because the root crown is located just below the soil surface, the plants are not easily killed by soil compaction or treading. The extensive underground root system enables the species to survive drought conditions. The plants overwinter below the ground if in open areas; they overwinter as small rosette plants if growing where they have cover, as in grasslands. Survival of seeds buried in soil is apparently due to induced dormancy that can only be broken by exposure to light. The longevity of seed buried in soil is greater than 20 years.

DISTINGUISHING CHARACTERISTICS

Broadleaf Plantain

The leaves are simple, basal, dark green, ovate, usually wavy-edged or angular toothed, 2 to 12 inches (5 to 30 cm) long and 1 to 4 inches (2.5 to 10 cm) wide, thick, roughish on one or both sides when dry, with minute hairs. The leaf blade has three to seven prominent ribs (veins) running from tip to petiole. The prominent petiole is usually one-third to two-thirds as long as the leaf blade, broad, usually green, and pubescent at the base.

The erect flower stalks are up to 20 inches (up to 50 cm) tall. The linear-cylindrical spikes are up to 10 inches (25 cm) long and 6 mm wide. There are 1 to 30 flower stalks per plant, and about 16 flowers per cm of spike.

The seeds are 1 to 1.7 mm long and 0.5 mm wide, light to dark brown, angled, and covered with fine ridges (reticulated). The capsules (seed pods) contain 5 to 22 seeds in each capsule. The capsules are nearly spherical (ovoid), about 5 mm long, and dehisce (split open) near the middle of the pod. Seed production is variable, with 565 seeds per plant recorded at London, Ontario. As many as 14,000 seeds per plant per season have been recorded.

Broadleaf plantain is distinguished by its long-petioled leaf blade, its many-seeded seed pod (capsule) that splits open above the base, and the angular seeds marked with wavy, threadlike ridges and a light-colored hilum.The leaves of broadleaf plantain have three to seven main veins; the bracts are obtuse-acute, 0.5 to 1.0 mm long, the sepals are 1.5 to 2.0 mm long; the capsule is rhombic-ovate, 4 to 5 mm long, dehiscent near the middle, with 5 to 22 seeds per capsule. Broadleaf plantain is often confused with blackseed plantain.

Buckhorn Plantain

Buckhorn plantain is one of the four most prevalent weeds in established legumes in the northeastern United States. Its airborne pollen, shed in large amounts, is an important contributor to summer hay fever. Buckhorn plantain is palatable to livestock and provides fodder of fair quality, although inferior to grasses. It is one of the most palatable species to sheep; they will actually chisel the crowns out of the ground using their lower incisors. It has been described as a particularly good source of calcium, iron, sodium, and other inorganic minerals. It is grown as a medicinal crop in Europe.

The leaves arise from the root crown (caudex) and are alternate in rosette. They are simple, basal, 3 to 16 inches (8 to 40 cm) long and 0.2 to 1.4 inches (0.5 to 3.5 cm) wide, lanceolate to lanceolate-oblong, entire or shallowly dentate, with 3 to 5 prominent ribs (veins) running from tip to petiole. The petiole is short. The leaves are glabrous or sparsely pubescent.

The erect flower stalks are up to 16 inches (40 cm) tall, often longer than the leaves. There are usually 3 or 4 flower stalks per plant. The ovoid-conical to cylindrical spikes are 0.8 to 3 inches (2 to 8 cm) long.

The seeds are 2 mm long, ellipsoid, brown or shiny black, concave on the inner face. The seed pods are ovoid-oblong, less than 6 mm long, contain one or two seeds per pod, and dehisce near the middle of the pod. Seed production is variable depending on time of season and location, averaging about 127 seeds per plant with an average of 2.6 spikes per plant in October at London, Ontario. At this same location near a deciduous woodland path, seed production averaged 60 seeds per plant with an average of 1.8 spikes per plant. In a rich arable field in the same area, individual plants growing without competition produced more than 10,000 seeds on more than 30 spikes.

Buckhorn plantain is distinguished by lanceolate leaves 0.2 to 1.4 inches (0.5 to 3.5 cm) wide with short petioles, the one- to two-seeded capsule which ruptures about the middle, the smooth concave seeds with a scar around the middle, and the dense, conelike spike with long, conspicuous, white stamens, about 7 mm long, jutting out from it. The leaf axils are often filled with long, brownish, cottony hairs. Buckhorn plantain is often confused with hoary plantain (*Plantago media*). However, hoary plantain contains four seeds per capsule, and has prostrate leaves that are elliptic to ovate and usually more than 3 cm wide.

Blackseed Plantain

Blackseed plantain is very similar morphologically to broadleaf plantain and is easily confused with it. However, they can be distinguished one from the other on the basis of leaf, bract, sepal, and capsule characters. The leaves of blackseed plantain have five to nine main ribs running from leaf tip to petiole; the sepals are sharply keeled, acute, and about 4 mm long. The leaves are 2.4 to 20 inches (6 to 50 cm) long, simple, erect or spreading, glabrous or inconspicuously pubescent, broadly elliptic to oval, wavy-edged or angular-toothed. The prominent petiole is one- to two-thirds as long as the leaf blade; its base is purplish.

The erect flower stalks are up to 20 inches (50 cm) tall. The spikes are up to 12 inches (30 cm) long and 6 mm wide. There are one to eight flower stalks per plant and about 10 flowers per cm of spike.

The seeds are 1.5 to 2.5 mm long, oval, angled, dark brown to black, with a scar (minute hilum) near the center, not reticulated. The brown or purple seed pods contain four to nine seeds per pod. The pods are nearly cylindrical, about 5 mm long, and split open across the middle of the capsule. Seed production is variable, with 662 seeds per plant recorded at London, Ontario.

REFERENCES

Cavers, P.B., I.J. Bassett, and C.W. Crompton. 1980. The biology of Canadian weeds. 47. *Plantago lanceolata* L. Can. J. Plant Sci. 60: 1269–1282.

Gaines, X.M., and D.G. Swan. 1972. Buckhorn plantain (*Plantago lanceolata*) and broadleaf plantain (*Plantago major*). In Weeds of Eastern Washington and Adjacent Areas. Davenport, WA: Camp-Na-Bor-Lee Association, pp. 252–255.

Hawthorne, W.R. 1974. The biology of Canadian weeds. 4. *Plantago major* and *P. rugelii*. Can. J. Plant Sci. 54: 383–396.

Holm, L.G., D.L. Plucknett, J.V. Pancho, and J.P. Herberger. 1977. *Plantago major* L. and *Plantago lanceolata* L. In The World's Worst Weeds. Honolulu: The University Press of Hawaii, pp. 385–393.

Korsmo, E. 1954. *Plantago lanceolata* L. In Anatomy of Weeds. Forlag, Norway: Grondahl and Sons, pp. 294–297.

Korsmo, E. 1954. *Plantago major* L. In Anatomy of Weeds. Forlag, Norway: Grondahl and Sons, pp. 298–301.

Murphy, T.R., D.L. Colvin, R. Dickens, J.W. Everest, D. Hall, and L.B. McCarty. 1993. Buckhorn plantain (*Plantago lanceolata*) and broadleaf plantain (*Plantago major*). In Weeds of Southern Turfgrasses. Athens, Georgia.Agricultural Business Office, University of Georgia, pp. 160–161.

Parker, K.F. 1972. Buckhorn plantain (*Plantago lanceolata* L.) and broadleaf plantain (*Plantago major*). Arizona Weeds. Tucson: The University Press of Arizona, pp. 268–271.

Reed, C.F., and R.O. Hughes. 1970. Buckhorn plantain (*Plantago lanceolata*) and broadleaf plantain (*Plantago major*). In Selected Weeds of the United States. Agriculture Handbook No. 366. Washington, DC: U.S. Department of Agriculture, U.S. Government Printing Office, pp. 268–269, 346–347.

Wax, L.M., R.S. Fawcett, and D. Isely. 1990. Buckhorn plantain (*Plantago lanceolata*) and broadleaf plantain (*Plantago major*). In Weeds of the North Central States, Bulletin 772. Urbana: University of Illinois at Urbana-Champaign, pp. 170–171.

Part Six

Perennial Broadleaved Weeds Reproducing from Buds on Creeping, Horizontal Roots

11

Canada Thistle
(Cirsium arvense)

Introduction

Canada thistle (Figure 11.1) is a herbaceous, perennial broadleaf weed. It is one of the most persistent and widespread weeds of the world. It infests cultivated fields, causing severe crop losses wherever sizable infestations occur. It also infests hay fields, pastures, rangelands, industrial sites, roadsides, stream banks, rights-of-way, parks, forests, and wastelands. It is a troublesome weed in gardens and lawns. Canada thistle is especially troublesome to people and livestock because of the plant's sharp spines. In the United States, Canada thistle is listed as a

Figure 11.1. Canada thistle plant, including part of extensive creeping, horizontal root system, A; Insets show details of a flower head, B; flower, C; and seed with and without hairy pappus, D.

Source: Reed, C.F., and R.O. Hughes. 1970. Canada thistle (*Cirsium arvense* L. Scop.). In *Selected Weeds of the United States.* Agricultural Handbook No. 336, p. 397. Washington, DC: U.S. Department of Agriculture, U.S. Government Printing Office.

prohibited noxious weed in the seed laws of 40 states and as a restricted noxious weed in nine states.

In Canada, *Cirsium arvense* is considered the typical variety of Canada thistle. However, the common variety of Canada thistle in Canada is *Cirsium horridum*. This variety is found in the agricultural areas of all the provinces, and it is the variety known to most people there. The leaves of *C. horridum* have a rather tough texture, stiff, the surface wavy, not flat, deeply and symmetrically lobed and pointed; marginal spines are numerous, stiff and stout, yellow, and longer than in other varieties. White-flowered plants of *C. horridum* have been collected in all provinces of Canada.

Distribution

Canada thistle is a common weed in the northern half of the United States (Figure 11.2). It was first introduced to North America as an impurity in crop seeds imported into Quebec and Ontario, Canada. It was recognized as early as the late 1700s as a problem weed in the United States. In 1900, Canada thistle was reported to be in all states of the United States north of the 37th parallel, roughly a line running east to west along the southern borders of Virginia, Missouri, Colorado, Utah, and continuing west to the coast through Fresno, California. Some of the most severe stands of this weed are found in Wisconsin, Minnesota, Idaho, Wyoming, and Washington.

Canada thistle is indigenous to Europe, western Asia, and northern Africa, and it is widely distributed in the temperate zones of South America, Australia, and New Zealand.

The Problem

Canada thistle is one of the most destructive and serious weeds in North America above the 37th parallel. It is found in all crop systems, and it seriously reduces yields of corn, peas, potatoes, soybeans, sugar beets, and forage crops. Perhaps the greatest losses occur in fields of spring-planted small grains, where the weed is ideally adapted to the cropping program. Canada thistle seeds are harvested along with crop seeds, reducing the commercial value of poorly cleaned legume, grass, and small grain seeds.

In the United States, the spread of Canada thistle was rapid. In

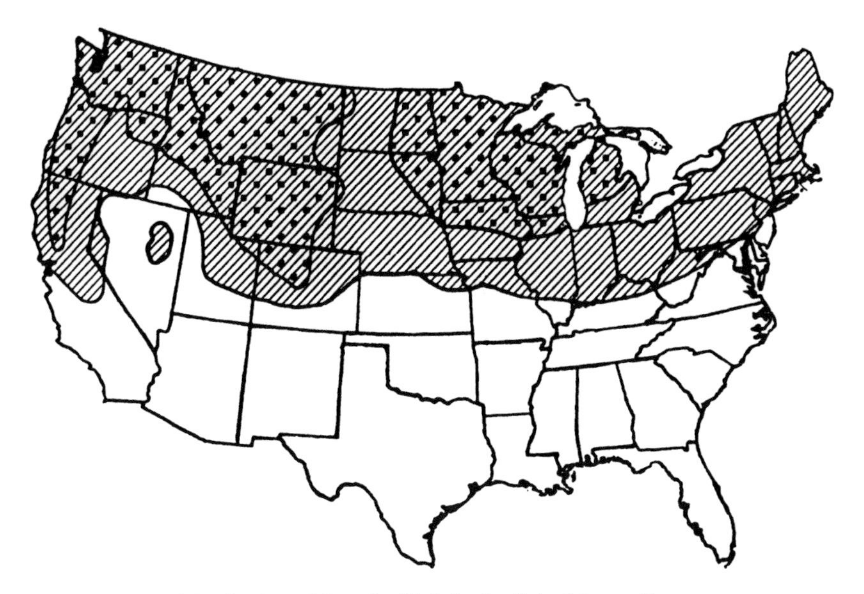

FIGURE 11.2. Distribution of Canada thistle in the United States. Denser cross-hatching denotes infestations of greater economic importance.

Source: Reed, C.F., and R.O. Hughes. 1970. Canada thistle (*Cirsium arvense* L. Scop.). In *Selected Weeds of the United States*. Agricultural Handbook No. 366, p. 396. Washington, DC: United States Department of Agriculture, U.S. Government Printing Office.

1901, Canada thistle was a rare weed in Montana, with only five small patches reported. By 1952, it was reported to infest more acreage than any other noxious weed in the four-state area of Montana, Idaho, Oregon, and Washington; by 1956, 625,000 acres (253,000 ha) were infested. It is reported to cause heavier losses in Idaho, Montana, and Washington than any other perennial weed. Canada thistle occurs in every county in Oregon and is the most serious, widespread, weed problem in the state. Canada thistle is found in all parts of Indiana, but it is most troublesome in the northeast part of the state. In Minnesota, Canada thistle is the second most troublesome weed in corn, reducing yields by 30 to 50%, especially in dry years. It is not a problem in the southern Corn Belt, as the weed needs the longer days and cooler temperatures of the north for aggressive growth.

In Montana, a survey comparing wheat yield reductions and Canada thistle shoot density found reductions of 15%, 35%, and 60% with stands of 2, 12, and 25 shoots per sq yd (0.8 sq m), respectively.

These shoot densities are not uncommon in wheat fields. Where only two shoots per sq yd were present, yield reduction per shoot was greater than at higher shoot densities because the Canada thistle plants were larger and more vigorous. In Idaho, a severe Canada thistle infestation reduced potato yields by 71%. Size and quality of the potatoes were adversely affected, and the weed vegetation interfered with harvesting.

PROPAGATION

Canada thistle reproduces by seed and adventitious buds on creeping roots and underground parts of aerial shoots; sometimes by underground crown buds on vertical stems of the parent plant.

SPREAD

Canada thistle spreads by seed dispersal, creeping roots, and by reproductive root and stem fragments being dragged by tillage tools to new locations. Canada thistle seeds are commonly spread as contaminants in crop seeds, and by flowing water in irrigation channels. The seeds may also be spread by most of the usual means by which other weed seeds are dispersed.

PERENNATION

The extensive vertical and horizontal root system, capable of asexual reproduction from adventitious buds, is responsible for the perennial nature of Canada thistle.

RHIZOMES

Canada thistle plants do not have rhizomes, despite this assertion in some weed identification guides.

ALLELOPATHY

It has not been proven conclusively that Canada thistle exhibits allelopathic properties, although reports in the literature strongly suggest this possibility.

Soil Type

Canada thistle grows in a wide range of soil types. It is well adapted to productive, deep, well-aerated soils where temperatures are moderate. It is less common or often absent from very light, dry soils. Canada thistle tolerates a soil with a 2% salt content. It grows under a wide range of moisture conditions but grows best under 16 to 30 inches of rainfall or under irrigation. Poorly aerated soils or high water tables limit its growth.

Distinguishing Characteristics

Canada thistle is distinguished from other thistles by its horizontal, branching roots, its erect stems branching toward the top, and formation of dense circular colonies, with each colony usually consisting of a single clone. Canada thistle is sometimes confused with common pasture thistles. However, the common pasture thistles are distinguished by a white cottony (woolly, cobwebby) material on their leaves and stems, and they have enlarged fleshy taproots rather than a horizontal root system.

Plant Description

Canada thistle plants are characterized by deeply penetrating, vertical roots and an extensive creeping, branching, horizontal root system. The roots often grow laterally 15 to 20 ft (4.5 to 6 m) in a single season. Horizontal and vertical roots can produce adventitious buds and roots at any point and these are capable of developing into new subterranean organs and new aerial shoots. The root buds lack an internally controlled dormancy period in fall and winter, and bud growth is limited by low temperatures. The root buds possess limited freezing tolerance, and their ability to survive in cold climates depends on their depth in the soil profile.

The roots contain an abundance of stored food reserves, and these reserves enable roots located below tillage depth to support the repeated initiation of new shoots for about one and one-half growing seasons, despite repeated cultivations that kill the shoots and prevent replenishment of food reserves. The normal low point of stored food reserves in roots is in June, when flower buds start to appear. Aerial shoots arise from adventitious buds on the main roots and on their branches and horizontal root runners; they are usually spaced about 8 to 12 inches (20 to 30 cm) apart.

The cotyledons of Canada thistle are *epigeous*, pulled aboveground by the elongating hypocotyl. They are dull green, relatively thick, about 1 cm long, and oblong to rounded-oval.

A sampling of Canada thistle roots found at soil depths to 21 inches (53 cm) showed that 84% of all roots were within 15 inches (38 cm) of the soil surface, with 54% in the upper 3- to 9-inch (8- to 233-cm) layer, 30% in the 9- to 15-inch (23- to 38-cm) layer, and 16% in the 15- to 21-inch (38- to 53-cm) layer. Although roots in some soils have penetrated as deep as 20 ft (6 m), most roots develop in the 3- to 12-inch (8- to 30-cm) layer below the soil surface.

Segmentation of Canada thistle roots and stems by tillage can result in dense stands of Canada thistle. These segments will be scattered into new areas and new plants will arise from these segments. Roots and underground stem crowns that are segmented during tillage possess sufficient food reserves to survive more than 100 days and produce new shoots and roots.

Ninety-five percent of root segments as small as 1.2 cm long and 3 to 6 mm in diameter can produce new plants. Longer segments containing more stored food can produce larger, more vigorous plants in a shorter time. New plants emerge from root and stem segments in about 15 days. However, when segmented at the end of the flowering period, only about 10% produced plants.

Canada thistle leaves are alternate, very crinkly, oblong or lance-shaped, usually dark green and with a grayish undersurface. Sharp spines are numerous on the outer edges of the leaves and on the branches and main stem of the plant. The leaves are deeply lobed, sessile, and clasping, with the length three to five times the width. Distinct leaf differences have been noted in Canada thistle ecotypes (Figure 11.3).

Canada thistle stems grow erect from 1 to 4 ft (30 to 120 cm), arising from numerous buds on the horizontal roots. Following mowing or grazing, shoots may also arise from lateral buds at nodes on the underground portion of the vertical stem of the parent shoot. Stems are slender, leafy, branching at leaf axils, and the main stem and branches are terminated by one to five sessile flower heads. Stems are usually green but may be brownish to reddish-purple in some clones or in some

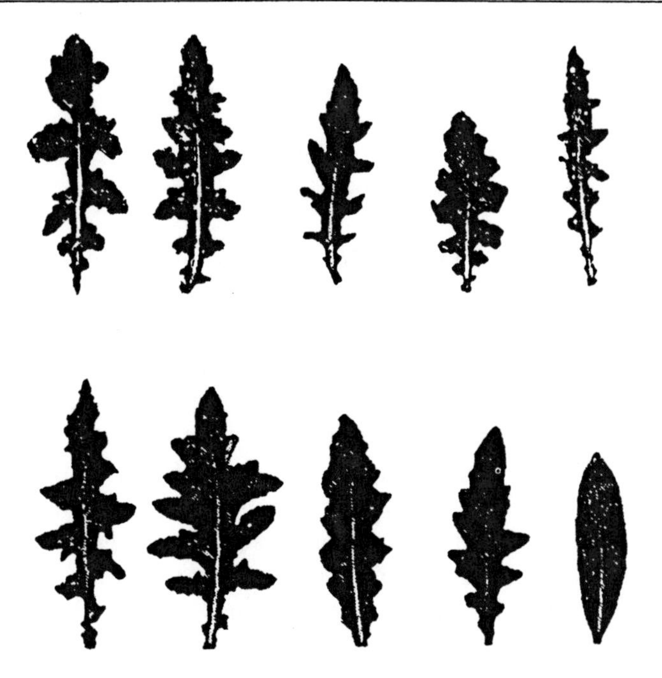

FIGURE 11.3. Typical leaves of 10 Canada thistle ecotypes in Wyoming, Montana, Idaho, and Washington.

Source: Agrichemical Age. 1974. p. 23.

locations. Stems of different clones vary from ribbed to smooth, glabrous to slightly pubescent, often with a row of spines below the leaves.

Canada thistle shoots normally emerge from the soil in May when average weekly temperatures are 42°F (5.6°C), but emergence is greatest when weekly temperatures average 46°F (7.8°C) or higher. Vertical growth begins and rapidly increases in about 3 weeks after shoot emergence, with an average growth of about 1 inch (2.5 cm) per day. In Colorado, new plants become fully developed, with flowers, in 7 to 8 weeks after shoot emergence in May. The aboveground vegetation of Canada thistle is killed by frost and freezing temperatures.

In a study with Canada thistle grown outdoors in boxes, it was found that, on an average, one Canada thistle plant produced 26 aerial shoots and 154 adventitious root buds after 18 weeks growth. A dense stand of Canada thistle may have 40 or 60 aerial shoots per 1.2 sq yd (1 sq m).

Canada thistle flowers are located in a "head" composed of about 100 flowers. Each plant has numerous fragrant flower heads located at the apex of the main stem and branches. Each compact flower head is 0.75 to 1.0 inch (2 to 2.5 cm) in diameter. The flowers are insect pollinated. The color of flowers on most plants are purple to rose; white flowers are found occasionally.

Canada thistle, a long-day plant, flowers from early July into August. Most ecotypes flower well when grown under 18 hours of daylight, but they do not flower when grown under 10 hours of daylight.

Canada thistle plants are dioecious (i.e., one plant has either male or female flowers, but not both). The flower types are easily identified in the field. The female flowers have a pappus of bristly hairs that is longer than the petals; the pappus of male flowers is shorter than the petals. The male flowers have abundant pollen on the staminate flowers; the female flowers have a voluminous pappus on the pistillate flowers. The plants are cross-pollinated by insects. A study in Iowa reported little seed production unless the male and female flowers are within 20 to 200 ft (6 to 60 m) of one another, a distance that can be worked by insects.

SEEDS

Canada thistle is a prolific seed producer, with quantities ranging from about 50 per flower head and 1500 to 5300 per plant. The seeds are oblong, light to dark brown, smooth, shiny, slightly tapered, about 0.1 to 0.14 inch (2.5 to 3.5 mm) long, more or less four-angled, flattened, curved or straight, with a ridge around the blossom end. A tuft of fine hairs (pappus) is attached to each seed. The pappus is white or tannish brown, plumose, and about 2 mm long. The seeds mature quickly after the flowers open and are capable of germinating in 8 to 10 days.

It is generally assumed that the pappus is the chief means by which Canada thistle seeds are dispersed long distances. However, the pappus often breaks free and drifts away while the seed is still in the flower head, leaving the seed in the head.

Seeds submerged in running water for 55 days germinated 26%; seeds suspended in mesh bags in flowing irrigation water for 22 months germinated 19%.

In general, currently used crop production systems which dispense with primary tillage (e.g., soil inversion by moldboard plowing) or employ reduced primary tillage (e.g., chisel plowing) tend to leave seeds on the soil surface or at depths of 1 to 3 inches (2.5 to 7.5 cm). Such a system favors germination of Canada thistle seed, resulting in seeds not being stored in the soil seed bank.

SEEDLINGS

Most Canada thistle seedlings emerge in April and May following soil burial the previous fall. Canada thistle seedlings are epigeous (i.e., the two cotyledons are pulled aboveground by the elongating hypocotyl). New seedlings are slow to become established and are quite sensitive to shading or competition from crops or other weeds. They often make a beginning on disturbed grazing areas, on open areas which are not grazed, or on noncrop areas such as ditchbanks.

REFERENCES

Anonymous. 1974. Controlling Canada thistle. Agrichemical Age 17(5): 23.

Anonymous. 1980. Canada thistle: Obnoxious in the north. Agrichemical Age 27(1): 9C.

Donald, W.W. 1990. Management and control of Canada thistle (*Cirsium arvense*). Rev. Weed Sci. 5: 193–250.

Donald, W.W. 1994. The biology of Canada thistle. Rev. Weed Sci. 6: 77–101.

Hayden, A. 1934. Distribution and reproduction of Canada thistle in Iowa. Amer. J. Bot. 21: 355–372.

Hodgson, J.M. 1964. Variation in ecotypes of Canada thistle. Weeds 12: 167–170.

Hodgson, J.M. 1968. The nature, ecology, and control of Canada thistle. Technical Bulletin No. 1386. Washington, DC: U.S. Department of Agriculture, U.S. Goverment Printing Office.

Hodgson, J.M. 1968. Canada thistle and its control. Leaflet No. 523. Washington, DC: U.S. Department of Agriculture, Superindentent of Documents.

Hodgson, J.M. 1974. Canada thistle: World-wide pest with many names. Weeds Today 5(1): 10–11.

Holm, L.G., D.L. Plucknett, J.V. Pancho, and J.P. Herberger. 1977. *Cirsium arvense* L. Scop. The World's Worst Weeds. Honolulu: University Press of Hawaii, pp. 217–224.

Korsmo, E. 1954. *Cirsium arvense* L. Scop. In Anatomy of Weeds. Forlag, Norway: Grondahl and Sons, pp. 350–355.

Miller, G.R. 1971. Today's weed: Canada thistle (*Cirsium arvense* L. Scop). Weeds Today 2(2):27.

More, R.J., and C. Frankton. 1974. The thistles of Canada. Monograph No. 10. Ottawa, Ontario: Research Branch, Canada Department of Agriculture.

Reed, C.F., and R.O. Hughes. 1970. Canada thistle. In Selected Weeds of the United States. Agricultural Handbook No. 366. Washington, DC: U.S. Department of Agriculture, Superintendent of Documents, pp. 396–397.

Sagar, G., and H. Rawson. 1964. The biology of *Cirsium arvense*. In J. Harper (ed.). Biology of Weeds. Oxford: Blackwell Scientific Publications, pp. 236–245.

Uva, R.H., J.C. Neal, and J.M. DiTomaso. 1997. Weeds of the Northeast. Ithaca, NY: Comstock Publishing Associates, Cornell University.

12

Common Milkweed
(Asclepias syriaca)

Introduction

Common milkweed (Figure 12.1) is a deep-rooted, creeping, herbaceous, perennial, broadleaved weed. It is found in cultivated lands, waste areas, river basins, and along fencerows, roadsides, and railroads. Stands of common milkweed may occur in soils of all textural groups, especially well-drained, loamy soils.

Common milkweed infests more than 26 million acres of land in the 13 north central states of the United States, with the greatest infestations in Iowa, Nebraska, and Wisconsin. Crops most affected are soybeans and corn, with 6 million and 12 million acres infested, respectively. Common milkweed is also a problem weed in peanuts, grain

Figure 12.1. Common milkweed. A. Plant habit ×0.5; B. flower, upper view ×3.5; side view ×2; C. seed pods (follicles) ×0.5; D. seeds with tufts of hairs (coma) ×3.

Source: Reed, C.F., and R.O. Hughes. 1970. Common milkweed (*Asclepias syriaca* L.). In *Selected Weeds of The United States*. Agriculture Handbook No. 366, p. 287. Washington, DC: U.S. Department of Agriculture, U.S. Government Printing Office.

sorghum, and fallow land. In addition to yield reductions, the floss attached to the seeds tends to clog air intake devices on combines during harvest. Common milkweed is also undesirable due to the unsightliness of the stout, tall plants.

In Canada, common milkweed is classified as a noxious weed in the Noxious Weeds Act of Ontario, Manitoba, and Quebec. It is considered a "class number one" noxious weed in the regulations to the Weed Control Act for Nova Scotia.

Common milkweed (and other milkweed species) contains several poisonous glucosidic substances (called cardenolides) known to be poisonous to sheep, cattle, and occasionally horses.

DISTRIBUTION

Common milkweed is native to North America. It is found throughout the eastern half of the United States, except for Louisiana, Texas, and

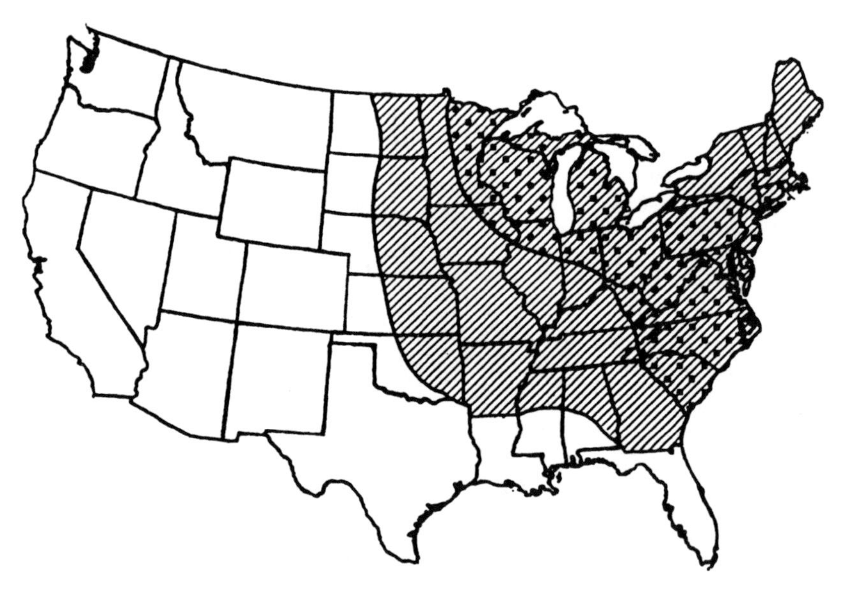

FIGURE 12.2. Distribution of common milkweed in the United States. Denser crosshatching denotes infestations of greater economic importance.

Source: Reed, C.F., and R.O. Hughes. 1970. Common milkweed (*Asclepias syriaca* L.). In *Selected Weeds of the United States*. Agriculture Handbook No. 366, p. 286. Washington, DC: U.S. Department of Agriculture, U.S. Government Printing Office.

portions of other states that border on the Gulf of Mexico (Figure 12.2).

In Canada, common milkweed is found in all provinces, from Saskatchewan eastward to the Atlantic, with the exception of Prince Edward Island.

RHIZOMES

Common milkweed plants do not have rhizomes, despite this assertion in some weed identification guides.

PROPAGATION

Common milkweed propagates by seeds and by adventitious buds formed on horizontal roots and/or the stem base (crown) near the soil surface.

SPREAD

Common milkweed seeds are dispersed by wind, aided by attached tufts of hairs. The seeds also float on water, and flowing water is another means of seed dispersal.

Established plants of common milkweed spread by creeping, horizontal roots, and an established plant may spread radially by as much as 10 ft (3 m) in a single growing season.

Cultivation is a major means by which established common milkweed plants are spread in cultivated fields. The cultivating tools cut or tear the roots into segments, and these segments are dragged or carried to new locations. Each root segment is capable of producing a new plant. In the field, root segments as short as 1 inch (2.5 cm) have produced shoots. In addition, the removal of aerial shoots by tillage or mowing induces sprouting of underground root buds, due to the loss of dominance induced by the shoot.

PERENNATION

Adventitious buds on horizontal roots and/or the stem base (crown) near the soil surface are the perennating organs responsible for the perennial habit of common milkweed.

Mistaken Identity

Showy milkweed (*Asclepias speciosa*) is very similar to common milkweed. However, it differs in having broad, oval and rounded or heart-shaped leaves, densely woolly stalks, and fewer and longer flowers.

Hemp dogbane (*Apocynum cannabinum*) may also be confused for common milkweed. In the field, hemp dogbane may be readily distinguished from common milkweed by its narrow, nearly sessile, elliptical leaves 2 to 5 inches (5 to 12 cm) long, by the prominent branching along the upper one-third or more of the stem, and when fruiting, by the long, dangling, narrow follicles (pods); refer to the description of hemp dogbane in Chapter 14.

Plant Description

Common milkweed characteristically grows as a colony of plants, and each colony develops from a single plant arising from a seed or root segment. All parts of the plant contain latex, a white, milky sap. Shoots from established plants arise from adventitious root buds, emerging from the soil in April and May. Root elongation occurs in July and August, ending in mid-August to mid-September. Shoots that develop from buds on these new roots do not reach the soil surface until the following spring.

The extensive root system is composed of both horizontal and vertical roots. Horizontal roots in the upper portion of the soil contain numerous adventitious buds capable of vegetative reproduction. In established stands, vertical roots may penetrate the soil to depths of 12.5 ft (3.8 m).

The stems of common milkweed are simple (unbranched), erect, stout, and 3 to 6 ft (1 to 2 m) high, with usually more than one stem arising from the same crown. The leaves are opposite, oblong, smooth margined, 4 to 10 inches (10 to 26 cm) long and 2 to 6 inches (5 to 15 cm) broad, with short, thick petioles. They are prominently veined, with the secondary veins nearly perpendicular to the midrib; the upper surface is green, smooth, and covered with a thick wax; the lower surface is covered with minute white hairs.

The small flowers have a sweet odor, are dull pinkish-purple, but many vary from nearly white to deep red in color. The flowers are arranged in large, showy, many-flowered clusters (umbels), round in shape, that arise from small, hairy pedicels at the top of the stems or in the axils of the upper leaves. The number of flowers per inflorescence varies greatly, from less than 10 to more than 120. However, only about 3% of the flowers produce seed pods, as most fall from the inflorescence in about 10 days after opening. The flowers are self-sterile; they are cross-pollinated by insects. Individual flowers are about 10 mm in diameter, and they are suspended by long, rather weak pedicels. The plants flower in June to August, depending on initial growth, climate, and location. The duration of flowering varies from plant to plant, umbel to umbel, and clone to clone.

The large seed pods (follicles) are grayish, hairy, and beset with soft spiny projections 1 to 3 mm high. There are usually four to six pods per stem, but there may be as many as 20 pods per stem. The pods are 3 to 4 inches (7.6 to 10 cm) long and 1 to 1.5 inches (2.5 to 3.8 cm) wide, and they taper to a curved point at the apex. The pods mature and split open along one side early in the fall (September-October), exposing their many seeds (150 to 425 seeds/pod). In Canada, an infestation of common milkweed was estimated to contain 24,300 stalks/A (60,000 stalks/ha), and the infested area, with an average of five seed pods/stalk and 290 seeds/pod, could produce 35 million seeds/A (87 million seeds/ha). The pods remain attached to dead stalks until late fall, or even until spring.

The seeds mature in August to October, and they are viable about 6 weeks after flowering. The seeds are brown, flat, and oval, about 7 mm long and 5 mm wide, with a tuft of silky white hairs (parachutes) attached at the tip. They may survive 3 years in soil.

A distinctive feature of common milkweed is the unique structure of its flower, common to the Milkweed family, in which the two ovaries and styles are covered (shut in) by the fusing of the five stamens with a broad stigmatic disk common to both pistils. Between the anthers of this enclosure are crevices (stigmatic chambers) into which the feet of nectar-seeking insects slip, and alternating with the crevices are cornucopialike projections containing nectar (nectar receptacles). On either side of a crevice is an anther sac, containing a coherent mass of pollen grains (pollinium; pl. pollinia). One pollen mass is joined to a similar mass in a contiguous anther by slender connectives across a crevice. Just above the crevice the two connectives are joined to a tiny dark body (a corpusculum) having a basal exterior slit tapering upward; this corpusculum stands just above the crevice between contiguous anthers into which a feeding insect's feet slip, providing a secure foot-hold while getting nectar from the nectar receptacles. As the insect leaves a flower its foot is drawn upward and gets caught in the V-shaped cleft of the corpusculum, and the two attached pollinia are then carried away by the insect and deposited in a crevice of another flower. The insect, on leaving the second flower, leaves the pollinia behind wedged in the crevice. Once the pollinia has been caught in a crevice, pollen tubes containing the sperms grow forth toward the pistils in the central cavity, penetrate the ovary cavity of each pistil and deposit the sperms; thereby fertilizing the eggs in the ovules. After fertilization, the pistils eventually become pods filled with seeds, each seed coming free with a tuft of hairs attached; the tuft being sloughed-off placental tissue (Stevens, 1948).

SEEDLINGS

Common milkweed seeds germinate at soil temperatures of 58° to 95°F (14.4° to 35°C). The seedlings emerge from soil depths of at least 2.5 inches (6.4 cm), regardless of soil type, and they readily become established in areas free of other plant competition. Common milkweed seedlings become perennial, capable of reproducing from its root system, approximately 3 weeks after seedling emergence. A lack of moisture is the limiting factor in seedling establishment.

Common milkweed seedlings grow normally under a long-day photoperiod (16 h) and a temperature of 81°F (27°C). Sixty-day-old seedlings may produce as many as 52 buds on their root system under these conditions. The rate of growth decreases as the temperature decreases to 59°F (15°C). Seedlings do not flower during the 1st year of growth. However, aerial shoots developing from the previous year's seedling root system flower normally.

Seedlings produce adventitious buds on the main root, located just below the soil surface, within 18 to 21 days after emergence. A study in Canada with one seedling, after 4 years of natural establishment, produced as many as 56 shoots vegetatively and 94 seedlings in an area of about 100 sq ft (9 sq m).

REFERENCES

Bhowmik, P.C. 1993. Biology and control of common milkweed (*Asclepias syriaca*). Rev. Weed Sci. 6: 227–250.

Bhowmik, P.C., and J.D. Bandeen. 1976. The biology of Canadian weeds, 19. *Asclepias syriaca* L. Can. J. Plant Sci. 56: 579–589.

Evetts, L.L., and O.C. Burnside. 1972. Germination and seedling development of common milkweed and other species. Weed Sci. 20: 371–378.

Evetts, L.L., and O.C. Burnside. 1974. Today's weed: Common Milkweed. Weeds Today 5(2): 19.

Jeffery, L.S., and L.R. Robison. 1971. Growth characteristics of common milkweed. Weed Sci. 19: 193–196.

Martin, A.R., and O.C. Burnside. 1980. Common milkweed: Weed on the increase. Weeds Today 11(1): 19–20.

Minshall, W.H. 1977. The biology of common milkweed. Proc. North Central Weed Control Conf. 32: 101–104.

Stevens, W.S. 1948. Milkweed family. In Kansas Wild Flowers. Lawrence: University of Kansas Press, pp. 177–178.

Uva, R.H., J.C. Neal, and J.M. Ditomaso. 1997. Weeds of the Northeast. Ithaca, NY: Comstock Publishing Associates, Cornell University Press.

Whitson, T.D. (ed). 1991. Weeds of the West. Laramie: University of Wyoming.

13

Field Bindweed
(Convolvulus arvensis)

Introduction

Field bindweed (Figure 13.1) is a prostrate or climbing, twining, herbaceous, perennial broadleaf weed. It is ranked 12th among the 18 worst weeds in the world. In the United States, it is listed as a prohibited noxious weed in the seed laws of 42 states and as a restricted noxious weed in seven states.

Field bindweed is a troublesome weed in both agronomic and horticultural crops. It is an aggressive weed in pastures, roadsides, open fields and edges of cultivated fields, gardens and yards, and waste areas. It climbs and becomes entangled on fences, plants, and other upright objects. It grows in dry and moderately moist soils, but it is not a weed of wetlands. Due to its aggressiveness and tenacity as a weed, field bindweed infestations "possess" the land, justifying one of its common names "*possession vine.*"

In California, field bindweed is considered one of the state's most serious weed problems, infesting 18% of the field crop acreage, and it is ranked 9th among the most troublesome weeds in cotton. In northern Oklahoma, field bindweed infests over 500,000 acres, and it is possibly the most serious weed problem in western Oklahoma, forcing some growers to give up crop production on normally very productive land. In North Dakota, field bindweed is found in every county, infesting about 3 million acres (1.2 million ha) of cropland. In Kansas during a 12-year period, field bindweed infestations reduced winter wheat yields an average of 30%.

Distribution

Field bindweed is native to Europe and was first reported in Virginia in 1739, soon spreading along the Atlantic seaboard and into New England. In about 1870, field bindweed was introduced into the Great Plains in wheat seed by German and Russian immigrants. It infests many agricultural areas of the United States, but it is best adapted to the semiarid regions west of the Mississippi River. It is one of the most harmful weeds in the Great Plains and western states. Field bindweed occurs throughout the United States, excepting the extreme Southeast, and parts of Texas, New Mexico, and Arizona (Figure 13.2).

FIGURE 13.1. Field bindweed. A. Habit ×0.5; B. root ×0.5; C. leaf variation ×0.5; D. flower, showing five stamens of unequal length ×1; E. capsule ×3; F. seeds ×4.

Source: Reed, C.F., and R.O. Hughes. 1970. Field bindweed (*Convolvulus arvensis* L.). In *Selected Weeds of the United States*. Agriculture Handbook No. 366, p. 291. Washington, DC: U.S. Department of Agriculture, U.S. Government Printing Office.

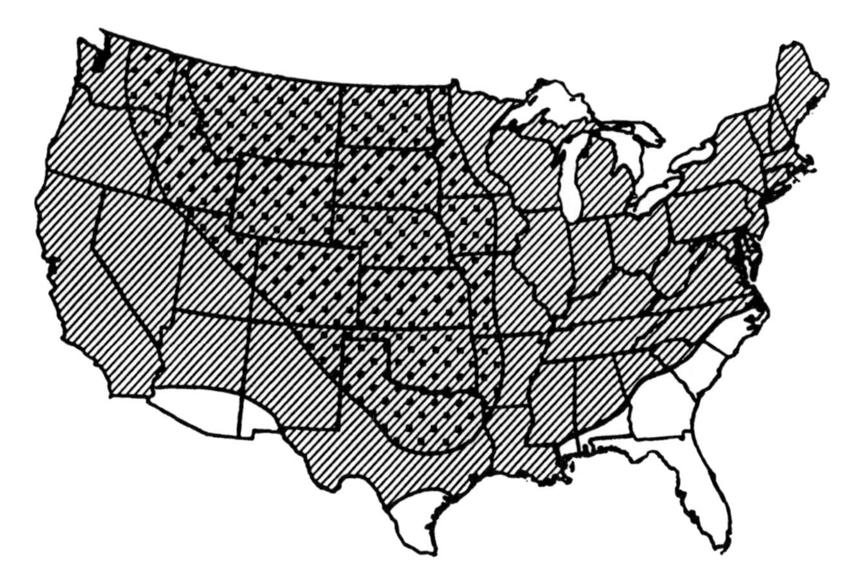

FIGURE 13.2. Distribution of field bindweed in the United States. Denser cross-hatching denotes infestations of greater economic importance.

Source: Reed, C.F., and R.O. Hughes. 1970. Field bindweed (*Convolvulus arvensis* L.). In *Selected Weeds of the United States*. Agriculture Handbook No. 366, p. 290. Washington, DC: U.S. Department of Agriculture, U.S. Printing Office.

In Canada, field bindweed is found in the southern parts of all provinces bordering the United States, from Nova Scotia to British Columbia.

RHIZOMES

Field bindweed plants do not have rhizomes, despite this assertion in some weed identification guides.

PROPAGATION

Field bindweed reproduces by seeds, adventitious buds on creeping roots, and axillary buds on belowground stem crowns.

SPREAD

Field bindweed spreads both by seeds and by an extensive creeping root system capable of asexual reproduction. Seeds are spread locally by dropping from seedpods on the spreading prostrate or vining stems. The seeds are spread more distantly when dropped from seedpods on vines wrapped around tillage tools and from harvesting equipment. Harvested with cereal crops, field bindweed seeds are difficult to separate from the crop seed, and they are often planted as a contaminant with crop seed. Field bindweed seeds ingested by livestock are largely undigested and, passing through the animals, are spread with manure.

Field bindweed spreads radially from the parent plant by a series of horizontal, creeping, lateral roots having adventitious root buds capable of producing new plants. The lateral roots of one plant can spread to 10 ft (61 cm) in diameter and produce 25 or more aerial shoots in one growing season. In two seasons, this radial spread can reach 18 ft (5.5 m) in diameter. Root and underground stem segments, with reproductive buds capable of producing new plants, are cut and spread by tillage equipment.

PERENNATION

The extensive root system and underground parts of vertical stems, both capable of asexual reproduction and overwintering, account for the perennial nature of field bindweed.

PLANT DESCRIPTION

Field bindweed is a low-growing, deep-rooted perennial weed with vigorous prostrate stems up to 10 ft (3 m) long, twining or spreading, but not rooting, over the surface of the ground. Once established, field bindweed is very difficult to control. It produces long-lived seed and persists vegetatively due to a deep regenerative root system. Although the aboveground vegetation of field bindweed is frost tolerant, this vegetation is killed by temperatures below freezing. The growing season for field bindweed extends from early spring to freezing temperatures in the fall.

The established root system of field bindweed consists of the primary vertical roots and many branching lateral roots. The vertical extensions become secondary vertical roots. Additional lateral roots develop at the bend where the preceding laterals have turned down. There are numerous reproductive buds both on the main roots and on the branches and horizontal roots. New shoots develop most freely at the bend of lateral roots and from underground vertical shoots (Figure 13.3).

The relatively shallow horizontal branching root system may cover an area 20 ft (6 m) in diameter. Numerous horizontal, lateral roots develop in the upper 2 ft (61 cm) of soil, with most in the upper foot of soil. When cultivated, the laterals tend to develop below the tilled area. About one-third of the total root system consists of the vertical roots below the 2-ft soil layer. Vertical roots grow an average of 15 ft (4.6 m) deep in three growing seasons, and they have been found as deep as 30 ft (9 m). Adventitious buds formed along lateral roots are capable of developing into shoots which, upon reaching the soil surface, become new plants.

The leaves of field bindweed are alternate, simple, very variable, petioled, up to 2 inches (5 cm) long and 1 inch (2.5 cm) wide, and arrowhead-shaped with two pointed basal lobes. Variation in leaf shape may be found on different plants of field bindweed, and to some extent on the same plant (Figure 13.4). The leaves of five biotypes of field bindweed are shown in the lower, right-hand corner of Figure 13.1.

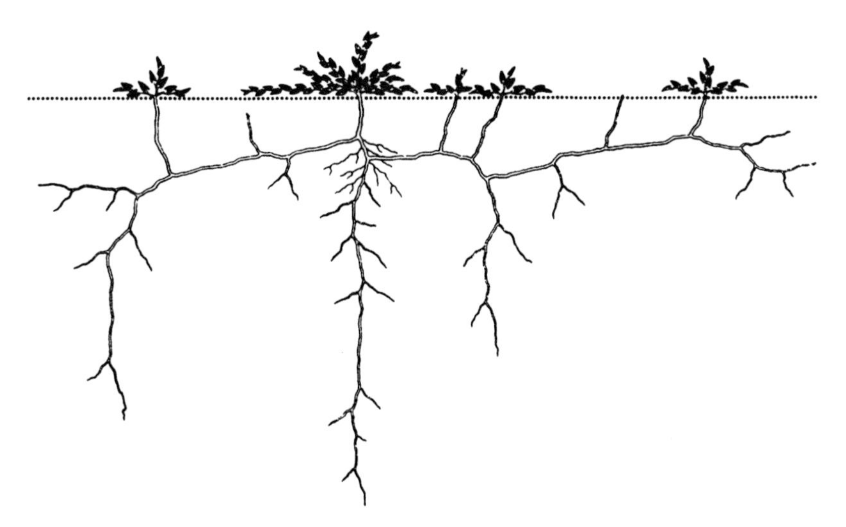

FIGURE 13.3. Diagram of the root system of field bindweed, showing the parent plant in the center with vertical and lateral roots; as the laterals turn downward, new shoots grow upward and form new plants.

Source: Swan, D.G. 1980. Field bindweed (*Convolvulus arvensis* L.). Bulletin 0888, 8 pp. Pullman: Washington State University. Illustrated by Hilary Broad. Reprinted by permission.

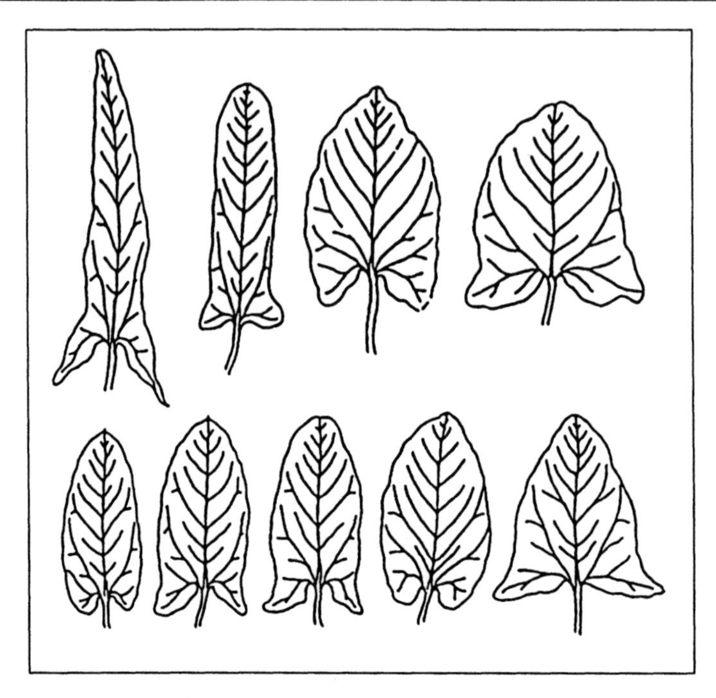

FIGURE 13.4. Nine leaf variations that may be found on different plants of field bindweed, and to some extent on the same plant.

Source: Degennaro, F.P., and S.C. Weller. 1984. Growth and reproductive characteristics of field bindweed (*Convolvulus arvensis* L.) biotypes. *Weed Science* 32: 525-528.

The flowers vary from white to pink. They are funnel-shaped, 0.75 to 1 inch (2 to 2.5 cm) wide and long, and occur singly (sometimes two or three) in the leaf axils on stalks (peduncles) 1 to 2 inches (1 to 5 cm) long. The flowers are similar in shape, though about one-half the size, to those of annual morning glory (*Ipomoea* spp.). A pair of narrow pointed bracts 0.1 to 0.2 inch (3 to 5 mm) long occur on the peduncles about 0.5 to 1 inch (1.2 to 2.5 cm) below the flowers. Sepals are bell-shaped, 0.1 inch (3 mm) long, oblong, and blunt. The plants flower from June to September in most regions, but from March to November in Arizona. Each flower produces a nearly round, point-tipped seedpod.

SEEDS

Each field bindweed seedpod usually contains four dark-brown to black seeds about 0.1 to 0.2 inch (3 to 5 mm) long. The seeds are somewhat egg-shaped (ovoid), the surface coarsely roughened, and three-sided. Two sides of the seed are flat and the third side is rounded. Seed is produced in abundance in hot and rather dry summers. In small grain crops, field bindweed produced about 650 seeds/plant. In North Dakota, soil samples revealed an estimated 2.7 million field bindweed seeds/A (1.1 million/ha) in the upper 6-inch (15-cm) soil layer.

The seed germinates in fall and spring but, unless the roots are sufficiently established, the seedlings are killed by temperatures of about 16°F (−9°C). Water, moist air, and fluctuating soil temperatures in the top 3 inches (76 cm) increase seed permeability. The following spring, about one-half of the seed, buried in soil the previous winter, germinate, but the remaining seed may not all germinate for many years. The optimum germination temperature is 86°F (30°C).

The cells of the seed coat mature about 1 month after pollination. When seed ripening reaches about 13% moisture, the cells of the seed coat thicken and fill with a substance that makes the seed coat impervious to water. Approximately 95% of the seed has a hard seed coat, impervious to water and water-soluble chemicals. The longevity of field bindweed seeds in soil is reported to be 20 years or more.

SEEDLINGS

Field bindweed seedlings have two notched, or heart-shaped, cotyledonary leaves (Figure 13.5). In contrast, young leaves on new shoots from established plants do not have notched leaves. The cotyledons are *epigeous,* pulled aboveground by the elongating hypocotyl. They are long-petioled, dark green, relatively large, square to kidney-shaped, and usually have a slight indentation at the apex. About 6 weeks after emergence, seedlings quickly develop a deep taproot and numerous lateral roots. Once lateral roots develop, the plants have a perennial growth habit. In one season, individual plants may spread to 10 ft (3 m) in diameter.

MISTAKEN IDENTITY

Hedge Bindweed

Field bindweed may be confused with hedge bindweed (*Calystegia sepium*; formerly, *Convolvulus sepium*) (Figure 13.6), a perennial, vining, climbing plant which is more common than field bindweed in the Corn Belt. The plants look similar, but hedge bindweed reproduces from axillary buds on rhizomes (Korsmo, 1954; Wells, 1972), whereas field bindweed reproduces from adventitious buds on horizontal roots. Also, hedge bindweed flowers lack bracts on the peduncle, which are present on peduncles of field bindweed flowers. The rhizomes of hedge bindweed are yellowish-white and relatively shallow.

Stems of hedge bindweed are 2 to 3 m long, smooth and prostrate. If during their growth they come in contact with other plants, they

FIGURE 13.5. Field bindweed seedling showing the two notched cotyledonary leaves.

Source: Swan, D.G. 1980. Field bindweed (*Convolvulus arvensis* L.). Bulletin 0888, 8 pp. Pullman: Washington State University. Photo by R.N. Harvey. Reprinted by permission.

FIGURE 13.6. Hedge bindweed. A. Plant habit ×0.5; B. root system ×0.5; C. diagram of flower ×0.25; D. seeds ×2.5.

Source: Reed, C.F., and R.O. Hughes. 1970. Hedge bindweed (*Convolvulus sepium* L.). In *Selected Weeds of the United States*. Agriculture Handbook No. 366, p. 293. Washington, DC: U.S. Department of Agriculture, U.S. Government Printing Office.

climb and twine around them in a counterclockwise direction. If the stem apex comes in contact with the ground, it grows down into the soil, becomes thicker, and forms a creeping rhizome. A stem lying on the soil surface puts forth fibrous roots that grow into the soil; new shoots subsequently form at these locations; thus, the stem may also serve as a stolon. The first two pairs of branches developed from the aerial shoot of the seed plant curve downward, penetrate into the soil, and form the first rhizomes of the plant.

The leaves, flowers, and bracts of hedge bindweed are much larger than those of field bindweed. The leaves are 2.5 to 3.5 inches (6 to 9 cm) long, flowers 1.5 to 2.5 inches (4 to 6 cm) long, and bracts 0.5 to 1 inch (1.3 to 2.5 cm) long. The leaves of hedge bindweed are more triangular and deeply lobed at the base than those of field bindweed.

Hedge bindweed spreads by creeping rhizomes, and by seeds dropped from flowers on slender stems climbing and twining on erect objects (e.g., fences, crop plants) or trailing (but not rooting) on the ground.

Hedge bindweed occurs throughout approximately the eastern half of the United States, extending as far west as Colorado and New Mexico, and at elevations of 6000 to 7000 feet in Arizona; throughout parts of the Pacific Northwest; north into southern Canada from British Columbia eastward to Newfoundland. The distribution of hedge bindweed in the United States is shown in Figure 13.7.

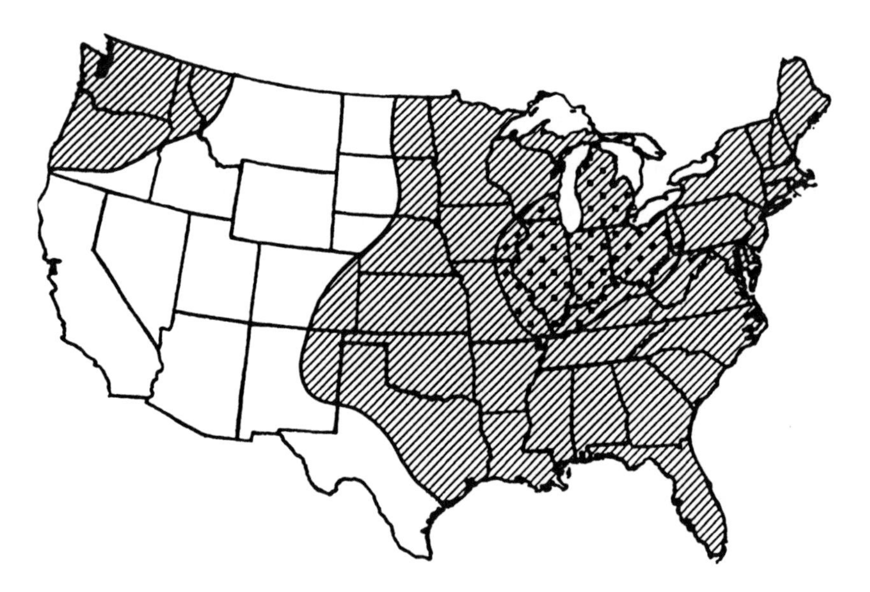

FIGURE 13.7. Distribution of hedge bindweed in the United States. Denser cross-hatching denotes infestations of greater economic importance.

Source: Reed, C.F., and R.O. Hughes. 1970. Hedge bindweed (*Convolvulus sepium* L.). In *Selected Weeds of the United States*. Agriculture Handbook No. 366, p. 292. Washington, DC: U.S. Department of Agriculture, U.S. Government Printing Office.

Wild Buckwheat

Field bindweed may be confused with wild buckwheat (*Polygonum convolvulus*) (Figure 13.8). Wild buckwheat is an annual, with smooth, slender, twining or creeping stems, branched at the base. Its leaves are alternate, heart-shaped, pointed with smooth edges. Flowers are small, greenish-white, and borne in clusters in leaf axils. Wild

FIGURE 13.8. Wild buckwheat. A. Plant habit ×0.5; B. branchlet with fruiting calyx ×4; C. flower ×5; D. achenes (seeds) ×5.

Source: Reed, C.F., and R.O. Hughes. 1970. Wild buckwheat (*Polygonum convolvulus* L.). In *Selected Weeds of the United States*. Agriculture Handbook No. 366, p. 121. Washington, DC: U.S. Department of Agriculture, U.S. Government Printing Office.

buckwheat may be distinguished from field bindweed by its annual habit, black, shiny three-cornered seeds, heart-shaped leaves and minute flowers. The distribution of wild buckwheat in the United States is shown in Figure 13.9.

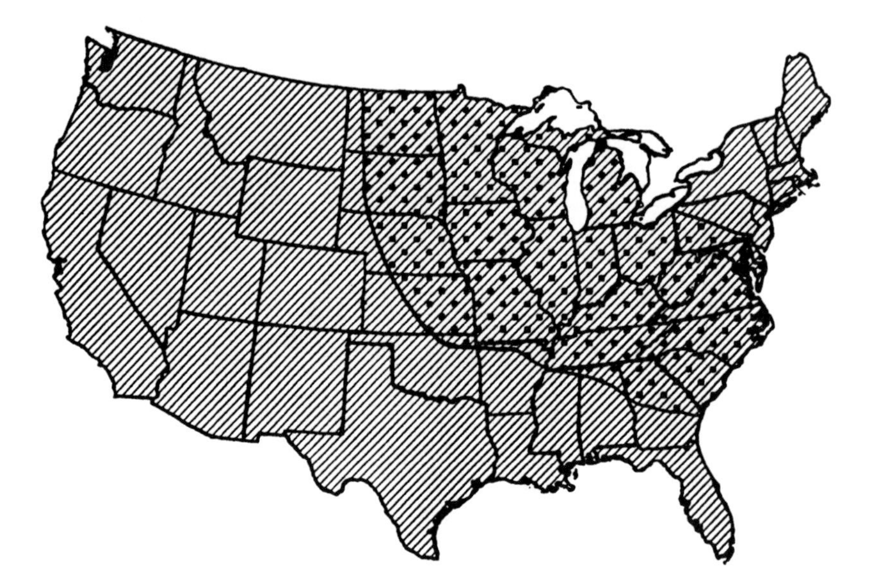

FIGURE 13.9. Distribution of wild buckwheat in the United States. Denser cross-hatching denotes infestations of greater economic importance.

Source: Reed, C.F., and R.O. Hughes. 1970. Wild buckwheat (*Polygonum convolvulus* L.). In *Selected Weeds of the United States*. Agriculture Handbook No. 366, p. 120. Washington, DC: U.S. Department of Agriculture, U.S. Government Printing Office.

REFERENCES

Anonymous. 1984. Field bindweed: Prevention and control. North Central Regional Extension Publication 206. Manhattan: Kansas State University, 7 pp.

Degennard, F.P., and S.C. Weller. 1984. Growth and reproductive characteristics of field bindweed (*Convolvulus arvensis*) biotypes. Weed Sci. 32: 525–528.

Holm, L.G., D.L. Plucknett, J.V. Pancho, and J.P. Herberger. 1977. *Convolvulus arvensis*. In The World's Worst Weeds. Honolulu: The University Press of Hawaii, pp. 98–104.

Korsmo, E. 1954. *Convolvulus sepium*. In Anatomy of Weeds. Forlag, Norway: Grondahl and Sons, pp. 260–263.

McWhorter, C.G., and J.R. Abernathy (eds). 1992. Weeds of cotton: Characterization and control. In No. 2, The Cotton Foundation Reference Book Series. Memphis, TN: The Cotton Foundation, p. 253.

Phillips, W.M. 1978. Special session on field bindweed; Field bindweed: The weed and the problem. Reprint No. 4. Reprinted from the Proceedings of the North Central Weed Control Conference 33: 140–148.

Reed, C.F., and R.O. Hughes. 1970. Field bindweed (*Convolvulus arvensis*). Selected Weeds of the United States. In Agriculture Handbook No. 366. Washington, DC: U.S. Department of Agriculture, U. S. Government Printing Office, pp. 290–291.

Reed C.F., and R.O. Hughes. 1970. Hedge bindweed (*Convolvulus arvensis*), Selected Weeds of the United States. In Agriculture Handbook No. 366. Washington, DC: U.S. Department of Agriculture, U.S. Government Printing Office, pp. 292–293.

Rosenthal, S.S. 1983. Field bindweed in California. Calif. Agri. 37(9–10): 16, 17, 22.

Swan, D.G. 1983. Regeneration of field bindweed (*Convolvulus arvensis*) seedlings. Weeds Today 14(4): 3–4.

Weaver, S.E., and W.R. Riley. 1982. The biology of Canadian weeds. 53. *Convolvulus arvensis* L. Can. J. Plant Sci. 62: 461–472.

Wells, W.A. 1972. In vitro studies of adventitious rooting in *Convolvulus sepium* L. Bot. Gaz. 133: 325–330.

Whitworth, J.W., and T.J. Muzik. 1967. Differential response of selected clones of bindweed to 2,4-D. Weeds 15: 275–280.

Wiese, A.F., and W.M. Phillips. 1976. Field bindweed. Weeds Today 6(4): 22–23.

14

Hemp Dogbane
(Apocynum cannabinum)

Introduction

Hemp dogbane (Figure 14.1), also known as Indian hemp, is a creeping, herbaceous, perennial broadleaf weed; woody at the base. It gets its name from the fact that native Americans used fiber from the bark for making rope.

Hemp dogbane can be a troublesome weed in cultivated crops. It grows in meadows, abandoned fields, gravelly or sandy fields, along roadsides, creekbeds and streams, irrigated ditches, and fence lines

FIGURE 14.1. Hemp dogbane. A. Plant habit ×0.5; B. rhizome ×0.5; C. flowers ×5; D. seed pods (follicles) ×0.5; E. seed with tuft of hairs (coma) ×1; F. seeds ×5.

Source: Reed, C.F., and R.O. Hughes. 1970. Hemp dogbane (*Apocynum cannabinum* L.). In *Selected Weeds of the United States*. Agriculture Handbook No. 366, p. 285. Washington, DC: U.S. Department of Agriculture, U.S. Government Printing Office.

along cultivated pastures. It is found in waste areas, dumps, thickets, and borders of woods. It grows on plains and foothills up to 7000 ft.

Hemp dogbane is a poisonous, herbaceous, perennial weed, and it may occasionally cause livestock losses in the western range states if ingested in sufficiently large amounts. However, hemp dogbane is extremely unpalatable to livestock due to its bitter, milky latex (sap). The leaves are poisonous at all times, even when dry. Normally, animals avoid hemp dogbane because of its bitter, sticky, milk-white sap. Sheep are more frequently affected than other animals, as they will eat large quantities of the leaves and tops if other forb-type plants are not available. Poisoning may also occur when sheep, cattle, and horses are trailed from summer to winter ranges. The chief toxic substance in hemp dogbane is thought to be *cymarin,* a glucoside once used as a heart stimulant for humans.

Distribution

Hemp dogbane is found in every state of the United States (Figure 14.2). In Canada, it occurs from western Quebec westward to Alberta.

Rhizomes

Hemp dogbane plants do not have rhizomes, despite this assertion in some weed identification guides.

Propagation

Hemp dogbane reproduces from seed and adventitious buds on long, creeping, horizontal roots and from crown buds along its woody base.

Spread

Hemp dogbane spreads by seeds and creeping roots.

Distinguishing Characteristics

Hemp dogbane grows in colonies. All plant parts contain latex, a milky sap, that exudes from cuts or breaks. The plants have a long, horizontal root system. The stems arise from a woody base. They are erect, 1

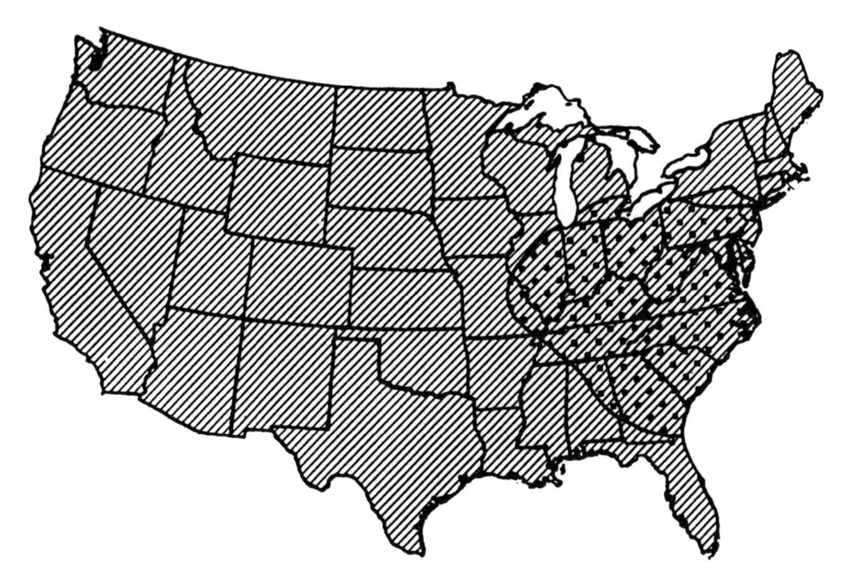

FIGURE 14.2. Distribution of hemp dogbane in the United States. Denser cross-hatching denotes infestations of greater economic importance.

Source: Reed, C.F., and R.O. Hughes. 1970. Hemp dogbane (*Apocynum cannabimun* L.). In *Selected Weeds of the United States*. Agricultural Handbook No. 366, p. 284. Washington, DC: United States Department of Agriculture, U.S. Government Printing Office.

to 5 ft (30 cm to 1.5 m) tall, and glabrous or nearly so. The upper portion of the stems has ascending branches, and the plant often appears bushy. The leaves are simple, opposite, erect or ascending, 2 to 6 inches (5 to 7.5 cm) long, ovate to lanceolate with a rounded wedge at the base and narrow to blunt point at the apex, smooth-edged, and glabrous to sparingly pubescent below. The leaves on the main stem have very short petioles 2 to 7 mm long and those on the branches are nearly sessile. The inflorescence (cyme) is at the tip of the stems, and the small flowers are mostly ascending, 2 to 4 mm long, with five greenish-white petals. After flowering, the pair of carpels develop into slender, cylindrical, brown, twined pods, at first united at their tips only, and filled with tufted seeds (Stevens, 1948). The prominent, slender, scythe-shaped seed pods (follicles) are in pairs for each flower, 6 to 8 inches (15 to 20 cm) long, hang down, and are dark reddish-brown in color. The seeds are thin and flat, 4 to 6 mm long, with a tuft of hair about 1 inch (2.5 cm) at the end. The plants flower from June to August, and the seed matures from August to October.

MISTAKEN IDENTITY

Similar-appearing dogbane species common to the western United States include prairie dogbane (*Apocynum sibericum*), spreading dogbane (*A. androsaemifolium*), and western dogbane (*A. medium*). Flower size is a key characteristic in distinguishing between the various dogbanes. Hybridization between species in this genus is reportedly frequent.

Common milkweed (*Asclepias syriaca*) may be mistaken for hemp dogbane due to its erect growth habit, thickish, light-green leaves, and milky sap. However, common milkweed leaves are more oblong and rounded, 1.5 to 4 inches (4 to 10 cm) long and 0.6 to 2.8 inches (1.6 to 7 cm) wide, stems not branched, flower clusters at tips of the stems and in the axils of the upper leaves. and seed pods (follicles) are upright, not slender and dangling. Refer to Chapter 12 for a description of common milkweed.

REFERENCES

Elmore, C.D. (Chrm.). 1985. Weed Identification Guide. Champaign, IL: Southern Weed Science Society.

James, L.F. et al. 1980. Plants Poisonous to Livestock in the Western States. In Agriculture Information Bulletin No. 415. Logan, UT: U.S. Department of Agriculture, Science and Education Administration, Poisonous Plant Research Laboratory.

Reed, C.F., and R.O. Hughes. 1970. Selected Weeds of the United States. In Agriculture Handbook No. 366. Washington, DC: U.S. Department of Agriculture, U.S. Government Printing Office.

Stevens, W.C. 1948. Kansas Wild Flowers. Lawrence: University of Kansas Press, p. 177.

Thornton, B.J., and H.D. Harrington. (No date given). Weeds of Colorado. Bulletin 514-S. Fort Collins, CO: Agricultural Experimental Station, Colorado State University.

Wax, L.M., R.S. Fawcett, and D. Isely. 1981 (Reprinted 1990). Weeds of the North Central States. Bulletin 772. Urbana: College of Agriculture, University of Illinois at Urbana-Champaign.

Whitson, T.D. (ed). 1991. Weeds of the West. Laramie, WY: Western Society of Weed Science, Bulletin Room, University of Wyoming.

15

Hoary Cress
(Cardaria draba)

Lens-podded Whitetop
(Cardaria chalepensis)

Globe-podded Whitetop
(Cardaria pubescens)

Introduction

There are three distinct species of *Cardaria* in North America, and four subspecies or varieties. The three species are hoary cress, also called whitetop (*Cardaria draba*) (Figure 15.1), lens-podded whitetop (*Car-*

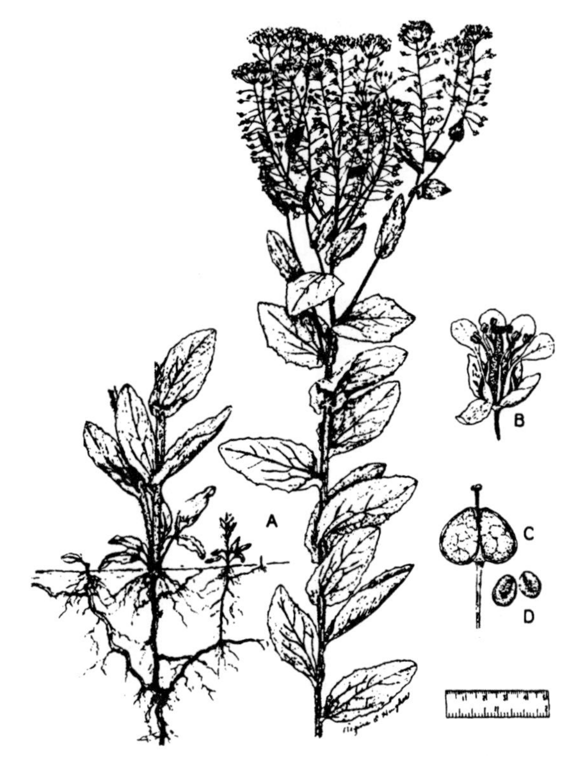

FIGURE 15.1. Hoary cress. A. Habit ×0.5; B. flower ×7.5; C. seedpod ×5; D. seeds ×7.5.

Source: Reed, C.F. and R.O. Hughes. 1970. Hoary cress (*Cardaria draba* L. Desv.). In *Selected Weeds of the United States*. Agriculture Handbook No. 366, p. 201, Washington, DC: U.S. Department of Agriculture, U.S. Government Printing Office.

daria chalepensis), and globe-podded whitetop (*Cardaria pubescens*). They are often confused for one another, as they are similar in general appearance and noxious characteristics.

In the United States, hoary cress and globe-podded whitetop are listed as prohibited noxious weeds in the seed laws of 19 and 13 states, respectively. Imported seed is prohibited entry into the United States if it contains seeds of any of the *Cardaria* species. In the United States, hoary cress is probably the weediest of the three *Cardaria* species. In Oregon, it is considered one of the most serious agricultural weeds. It is common throughout most of the counties east of the Cascade Mountains and is occasionally found in the western counties. Lens-podded whitetop is more widespread and troublesome in California than either hoary cress or globe-podded whitetop.

In Canada, all three *Cardaria* species are listed under the Seeds Act and Regulations administered by Agricultural Canada (1967) in the category, Prohibited Noxious Weed Seeds, Class 1. They are also listed in the Noxious Weeds Acts of Manitoba, Saskatchewan, Alberta, and British Columbia. Lens-podded whitetop is the most serious of the three *Cardaria* species, especially in irrigated areas of Saskatchewan, Alberta, and British Columbia. Hoary cress is the least weedy of the three *Cardaria* species. It is most troublesome in the Prairie Provinces, especially southern Manitoba and southern Alberta. Globe-podded whitetop grows in the same areas as hoary cress and lens-podded whitetop, but is usually less aggressive.

All three *Cardaria* species are a significant hazard to crop production under moist conditions and on irrigated land, but they are unlikely to become a major problem under drier conditions. They grow under open, unshaded conditions, and they are found in hayfields (especially alfalfa and brome), grainfields, grasslands, meadows, gardens, feed lots, along irrigation ditches and other watercourses, roadsides, and waste areas.

DISTRIBUTION

All three *Cardaria* species are native to Central Europe and the Mediterranean-Caspian Sea area. Hoary cress is naturalized throughout Europe.

Hoary cress was first reported in the United States in 1862 on Long Island, New York, and it was probably introduced in soil used as ships'

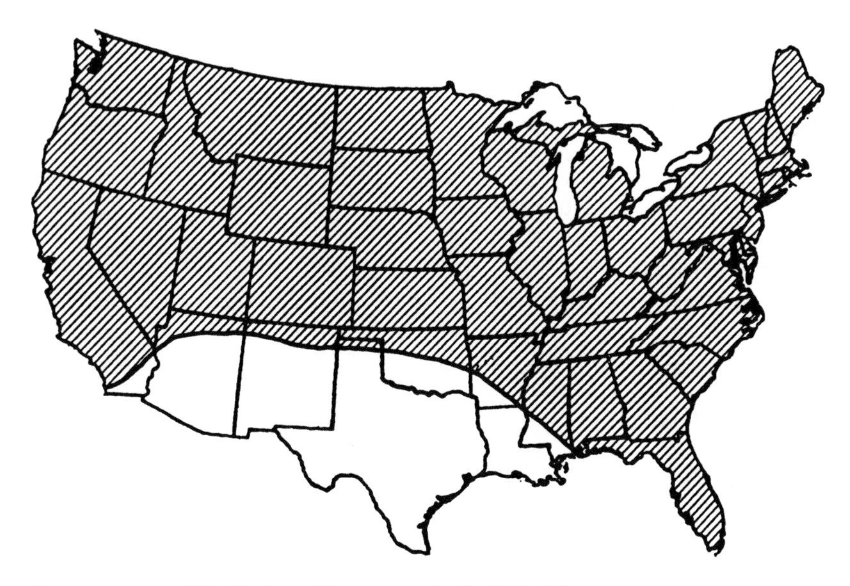

FIGURE 15.2. Distribution of hoary cress in the United States.

Source: Reed, C.F., and R.O. Hughes. 1970. Hoary cress (*Cardaria draba* L. Desv.). In *Selected Weeds of the United States*. Agriculture Handbook No. 366, p. 200. Washington, DC: U.S. Department of Agriculture, U.S. Government Printing Office.

ballast. It is now present as a weed in much of the United States (Figure 15.2). Hoary cress was first reported in Canada in 1878, at Barrie, Ontario. It is now found in all provinces of Canada, except for Newfoundland, Prince Edward Island, and New Brunswick.

Lens-podded whitetop and globe-podded whitetop are common, aggressive weeds in central and western North America; they are rare in eastern North America. Both species were introduced into the United States and Canada in alfalfa seed imported from Turkestan. Lens-podded whitetop was first reported in the United States in 1918, at Chino, California, and in Canada in 1926 at Grande Prairie, Alberta. Globe-podded whitetop was first reported in the United States in 1919, at Ypsilanti, Michigan, and in Canada in 1926, at Grande Prairie, Alberta.

RHIZOMES

The *Cardaria* species do not have rhizomes.

PROPAGATION

The *Cardaria* species reproduce by seeds and by adventitious buds on roots.

SPREAD

The *Cardaria* species spread by seed dispersal and by creeping roots and dispersal of root segments by tillage tools and other farm machinery.

PERENNATION

The extremely persistent, reproductive roots, with abundant food reserves, are responsible for the perennial nature of the three *Cardaria* species, allowing them to overwinter and produce new shoots in the spring.

DISTINGUISHING CHARACTERISTICS

Cardaria species are members of the mustard family (*Brassicaceae,* formerly *Cruciferae*), but they differ from all other mustards in that they have white flowers, leaf surfaces covered with short simple hairs, upper leaves clasping stem by basal lobes, and spread by creeping roots.

The three species are distinguished from one another primarily by the shape of their seedpods. The seedpods of hoary cress are heart shaped; those of lens-podded whitetop are nearly round or kidney shaped and compressed perpendicular to the partition, and those of globe-podded whitetop are globular to ellipsoid. Globe-podded whitetop has short simple hairs on the sepals and pods, whereas the other two *Cardaria* species have glabrous (without hairs) sepals and pods.

PLANT DESCRIPTION

Hoary cress is a herbaceous, perennial broadleaf plant. Its stems are erect or spreading, branched, sparsely pubescent, and 0.5 to 2 ft (15 to 60 cm) tall. The leaves are alternate, grayish-green, simple, mostly toothed. The basal leaves are 1.5 to 4 inches (4 to 10 cm) long, petioled, and ovate-lanceolate in shape. The upper leaves are entire, sessile (without petioles) with basal lobes clasping the stem, 1 to 2.5 inches (2 to 6.5 cm) long, and oblong or tapering to a point.

The flowers are small and numerous, clustered in terminal, flat-topped, corymblike racemes, with four white petals 3 to 5 mm long. They are cross-pollinated by insects. The plants flower in April to June, depending on location. In Oregon and Canada, they flower in early to mid-May. The plants are showy with their numerous white blossoms in flat-topped clusters, and infested areas in full bloom often appear almost pure white.

The seedpods are heart-shaped, glabrous, 2 to 5 mm long, and, at maturity, contain two (rarely four) reddish-brown, wingless, seeds about 2 mm long. The seeds are about the same size as alfalfa seed, and are difficult to remove from alfalfa and similar crop seeds. One plant of hoary cress can produce as many as 850 mature seedpods and 1200 to 4800 seeds per flowering stem. The germination of hoary cress seed after 1, 2, and 3 years was 84%, 31%, and 0%, respectively.

Hoary cress cotyledons appear aboveground 5 to 6 weeks after seed germination. During the next 3 weeks, the first leaves emerge and form a loose rosette. Lateral roots develop from the radicle in about 2 to 3 weeks. By 25 days of age, single plants have developed a taproot to a depth of 10 inches (25 cm), five or six horizontal lateral roots with numerous vegetative buds, and several shoots. After 100 days, single plants may develop as many as 48 aerial shoots spread over an area 2 ft (61 cm) in diameter. In its 1st year, a single plant of hoary cress established in the absence of competition may spread over an area 12 ft (3.7 m) in diameter and produce 455 aerial shoots. In Saskatchewan, radial increases ranged from 4 to 5 feet (61 to 71 cm) annually.

Lens-podded whitetop is similar in most respects to hoary cress. The stems are stoutish, erect to spreading, 8 to 16 inches (20 to 40 cm) tall, glabrous above and scattered to densely pubescent below. The basal leaves are irregularly toothed to entire, narrowed to a petiole, with sparse to dense pubescence. Middle and upper leaves are sessile, with two lobes clasping the stem, glabrous or sparsely pubescent, elliptic or lanceolate, and up to 3 inches (8 cm) long and 1.2 inches (3 cm) broad.

The seedpods are transversely lens-shaped, glabrous, inflated, 2.5 to 6.0 mm long and 4.0 to 6.0 mm broad. The plants can produce as many as 850 mature seedpods per flowering stem. The pods contain one to four seeds/pod, but are sometimes seedless. One and 3 years after maturity, seed germination was 86% and 52%, respectively. In California, lens-podded whitetop comes into bloom later than hoary cress and remains in flower for a longer period.

Globe-podded whitetop is similar in most respects to hoary cress and lens-podded whitetop. The stems are stoutish, erect to spreading, 4 to 16 inches (10 to 40 cm) tall, and sparsely to densely pubescent. The basal leaves taper to a petiole. The middle to upper leaves are sessile, with two lobes clasping the stem, linear-oblong to lanceolate, and 3inches (8 cm) long and 0.8 inch (2 cm) broad, but usually smaller. The inflorescence forms a compact corymb, with both long and short racemes. The petals are white and 2 to 3 mm long. In California, globe-podded whitetop starts blooming at the same time as lens-podded whitetop but remains in bloom well into winter.

The seedpods are sparsely to densely pubescent, greatly inflated, 3.0 to 4.5 mm long and 2.5 to 4.5 mm broad. Mature seedpods are frequently seedless, or with one or two, rarely three or four, seeds per pod. The seeds are oval, slightly compressed, reddish-brown in color, and 1 to 1.5 mm long by 1.5 to 2 mm broad. Mature plants of globe-podded whitetop produce an average of 300 pods, ranging from 30 to 560 per plant.

In Saskatchewan, seedlings of globe-podded whitetop emerge about May 1 and produce basal rosettes. By May 18, the seedlings were 2 to 5 inches (5 to 13 cm) tall, with five to seven leaves. By June 3, the plants were 5 to 7 inches (13 to 18 cm) tall, had 12 leaves, and were initiating flower buds. By July 7, some flowers were open, and by July 14 flowers were replaced by seedpods, some fully developed. On July 20, seeds were fully developed, but immature.

In Alberta, globe-podded whitetop flowers 10 to 14 days later than either hoary cress or lens-podded whitetop. It also blooms later than the other two species in Nevada. In western Canada under comparable conditions, when hoary cress and lens-podded whitetop are in flower, globe-podded whitetop is still in the early bud to flowering stage.

ROOTS

The following characteristics are common to hoary cress, lens-podded whitetop, and globe-podded whitetop: The root systems consist of a vertical taproot that develops one or more lateral (first-order) roots from which new rosettes and flowering shoots arise. These roots eventually turn down to become secondary vertical roots that often reach greater depths in the soil than the parent root. The vertical roots of hoary cress frequently penetrate the soil to depths of 29 to 32 inches (74 to 81 cm), occasionally as much as 5 ft (1.6 m). Roots of globe-podded whitetop have been traced as deep as 18.5 ft (5.6 m). Deep vertical penetration of the roots makes it difficult to easily control these species by cultivation. Lateral roots of the second order are developed on first-order laterals usually just below the point where the latter tend to become vertical (compare this location with that from which aerial shoots arise, mentioned below). This method of development is repeated until laterals of the third, fourth, and higher orders are formed.

The roots store a maximum of carbohydrates (chiefly in the form of starch) by about August 1. The lowest carbohydrate content occurs in spring to early summer when food reserves are being used to support new growth. In roots of undisturbed plants, the stored starch content can be more than 21% of the fresh weight; the total sugar content does not exceed 7.4%; and reducing sugars reach a maximum of 1.4%.

Rosettes and shoots develop from adventitious buds, which can form on any part of the permanent root system. Such buds give rise directly to new rosettes if borne at or near the soil surface. Aerial shoots tend to arise most often at or just above the point where lateral roots bend down to become vertical (compare this location with that from which lateral roots arise, mentioned above).

REFERENCES

Anonymous. 1956. Whitetop (*Cardaria draba*). Leaflet 49. Bozeman: Extension Service, Montana State College.

Hawkes, R.B., T.D. Whitson, and L.J. Dennis. 1985. Hoary cress (*Cardaria draba* L. Desv.). In A Guide to Selected Weeds of Oregon. Salem: Oregon Department of Agriculture, p. 32.

Meister, R.T. (ed). 1996. Weed Control Manual. Willoughby, OH: Meister Publishing Co. 538 pp.

Mulligan, G.A. and J.N. Findlay. 1974. The biology of Canadian weeds. 3. *Cardaria draba*, *C. chalepensis*, and *C. pubescens*. Can. J. Plant Sci. 54: 149–160.

Reed, C.F., and R.O. Hughes. 1970. Hoary cress (*Cardaria draba* L. Desv.). In Selected Weeds of the United States. Agricultural Handbook, No. 366.

Washington, DC: U.S. Department of Agriculture, U.S. Government Printing Office, pp. 200–201.

Robbins, W.W., M.K. Belleau, and W.S. Ball. 1951. Whitetop or hoary cress (*Cardaria draba* Desv.) In Weeds of California. Sacramento: State Department of California, pp. 220–222.

16

Horsenettle
(Solanum carolinense)

Silverleaf Nightshade
(Solanum elaeagnifolium)

Introduction

Horsenettle, also called Carolina nettle (Figure 16.1), and silverleaf nightshade, also called white horsenettle (Figure 16.2), are creeping, herbaceous, perennial broadleaf weeds. The plants of both species appear similar to the casual observer, and they are often mistaken for one another.

Horsenettle is listed as a prohibited noxious weed in the seed laws

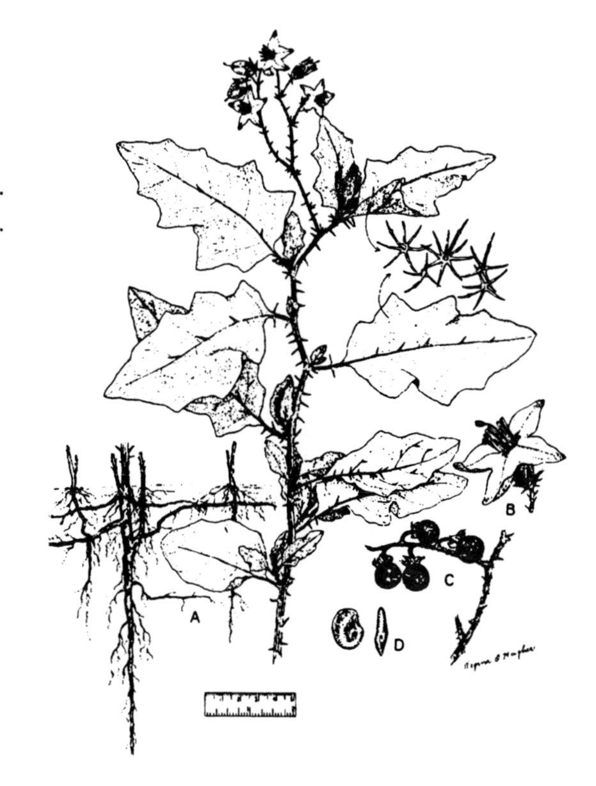

FIGURE 16.1. Horsenettle. A. Plant habit ×0.5; B. flower, ×1.5; C. berries ×0.5; D. seeds ×5.

Source: Reed, C.F., and R.O. Hughes. 1970. Horsenettle (*Solanum carolinense* L.). In *Selected Weeds of the United States.* Agriculture Handbook No. 366, p. 323. Washington, DC: U.S. Department of Agriculture, U.S. Government Printing Office.

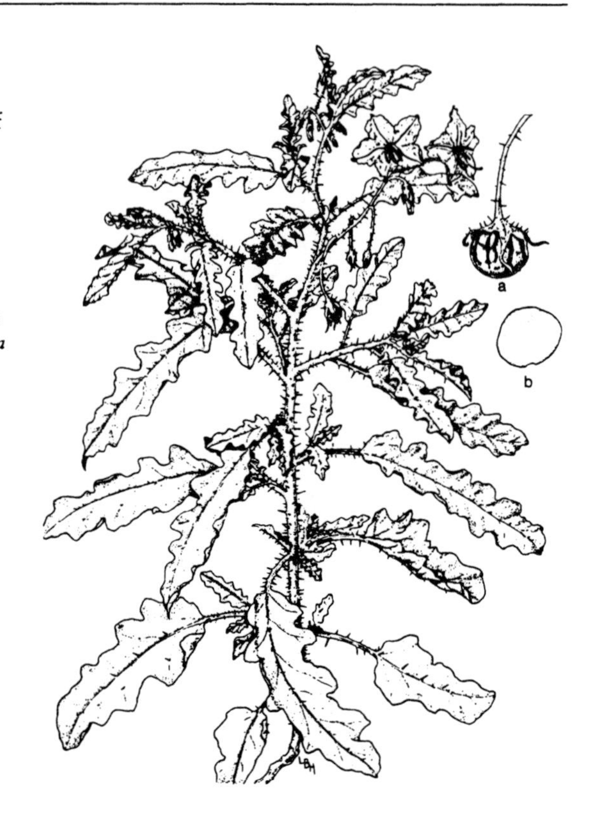

FIGURE 16.2. Silverleaf nightshade. Spiny plant with flowers and seedpods (berries); a, immature, striped seedpod; b, seed profile.

Source: Illustrations by Lucretia B. Hamilton from *An Illustrated Guide to Arizona Weeds* by Kittie Parker. Copyright ©1972 The Arizona Board of Regents. Reprinted by permission of the Univerity of Arizona Press.

of 11 states of the United States, and as a restricted noxious weed in 25 states. In Canada, horsenettle is listed as a noxious weed in Manitoba, but not in any other Canadian province.

Silverleaf nightshade is listed as a prohibited noxious weed in eight states of the United States, and as a restricted noxious weed in 11 states.

Plants of both *Solanum* species are found in cultivated lands, orchards, vineyards, pastures, rangelands, ditchbanks, roadsides, and waste areas. They generally grow best in moist sandy or gravelly soils, but they can thrive in clay and clay pan soils. They grow rapidly during hot weather, and they are very drought resistant.

DISTRIBUTION

Horsenettle is native to the southern United States. It is established throughout the eastern half of the United States, except in Maine (Figure 16.3). It is found in the western United States from Oregon and

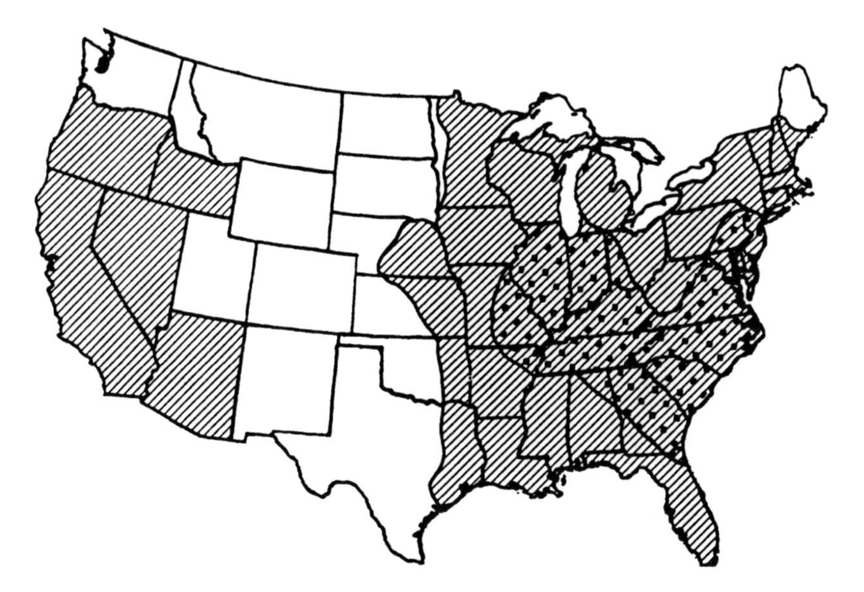

FIGURE 16.3. Distribution of horsenettle in the United States. Denser cross-hatching denotes infestations of greater economic importance.

Source: Reed, C.F., and R.O. Hughes. 1970. Horsenettle (*Solanum carolinenese* L.). In *Selected Weeds of the United States*. Agricultural Handbook No. 366, p. 322, Washington, DC: United States Department of Agriculture, U.S. Government Printing Office.

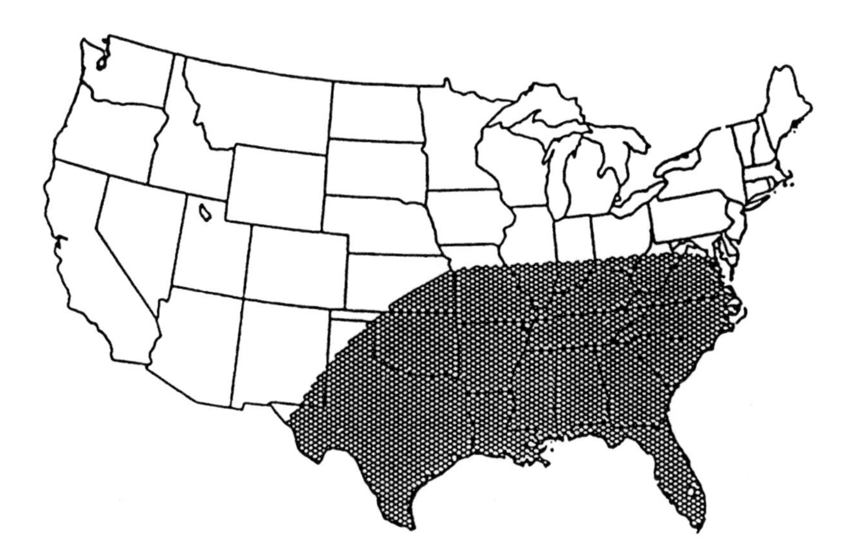

FIGURE 16.4. Distribution of silverleaf nightshade in the United States.

Source: Elmore, D.C. (chrm.) 1985. *Weed Identification Guide*. Champaign, IL: Southern Weed Science Society. Reprinted by permission.

California east to Arizona and to the Rockies in southern Idaho, and locally in New Mexico and other areas.

Silverleaf nightshade is a native to the central United States, but it has spread to other areas of the country (Figure 16.4). Silverleaf nightshade has not been reported in Canada.

RHIZOMES

Horsenettle and silverleaf nightshade do not have rhizomes, despite this assertion in some weed identification guides.

PROPAGATION

Both *Solanum* species propagate by seeds and by adventitious buds on creeping horizontal roots and crown buds located at or below the soil surface.

SPREAD

Both *Solanum* species spread by seeds and by creeping roots.

PERENNATION

Adventitious buds on deep-seated, creeping, horizontal roots and crown buds located at or below the soil surface are responsible for the perennial habit of both horsenettle and silverleaf nightshade.

DISTINGUISHING FEATURES

The flowers of horsenettle are white, pale blue, or violet to whitish; those of silverleaf nightshade are violet or blue. Horsenettle stems and leaves lack a dense, silvery covering of hairs; whereas, the entire silverleaf nightshade plant, except for the upper surface of the leaves, appears whitish, due to a covering of very, fine, whitish, star-shaped hairs. Plants of horsenettle are generally more prickly than those of silverleaf nightshade, with the branches, stems, and leaf mid-ribs, and sometimes even the lateral leaf veins armed with straight yellow spines. A special identifying feature of silverleaf nightshade is that the upper surface of each leaf is green and the under surface is silvery-green.

In the fall, after the aboveground vegetation of both species is killed by frost or freezing temperatures, the leaves drop from the main stems and branches, and the persistent, colorful berries (pods) are easily visible from a distance, enabling one to make a quick, general identification.

Root System

Both species have an extensive root system with taproots, fibrous feeder roots, and creeping, horizontal roots. Vertical taproots penetrate the soil to depths of 10 ft (3 m) or more. The shallower, creeping, horizontal roots develop as lateral roots from the taproots. They are found in the upper 18-inch (45-cm) soil layer and extend 3 or 4 ft (1 to 1.2 m) or more from the taproot. Both the vertical and horizontal roots are capable of producing shoots from adventitious buds. Roots exposed to freezing temperatures in the upper 2.4-inch (6-cm) soil layer are susceptible to kill.

A 6-inch (15-cm)-long segment cut from a vertical root of horsenettle produced four new aerial shoots when planted 18 inches (46 cm) below the soil surface. New horsenettle plants have been produced from root segments as short as 2 mm. Figure 16.5 illustrates a horsenettle plant developing from an adventitious bud on a severed root segment, 2 and 6 weeks after emergence.

Stems

The plants of both species have stems that are about 1 to 4 ft (30 to 120 cm) tall, erect, spiny or spineless, leafy, and loosely branched.

Leaves

The leaves of horsenettle are green, simple, alternate, 2 to 6 inches (5 to 15 cm) long (including petioles) and 1 to 3 inches (2.5 to 7.5 cm) wide, oblong to ovate, unevenly lobed, wavy margins, with yellow prickles on petioles, midrib, and veins. The upper and lower surfaces are sparsely covered with fine, stiff, but not white, star-shaped, four- to eight-rayed hairs.

The leaves of silverleaf nightshade are simple, alternate, 1.5 to 4 inches (4 to 10 cm) long (including petioles), and 0.25 to 1 inch (0.6 to

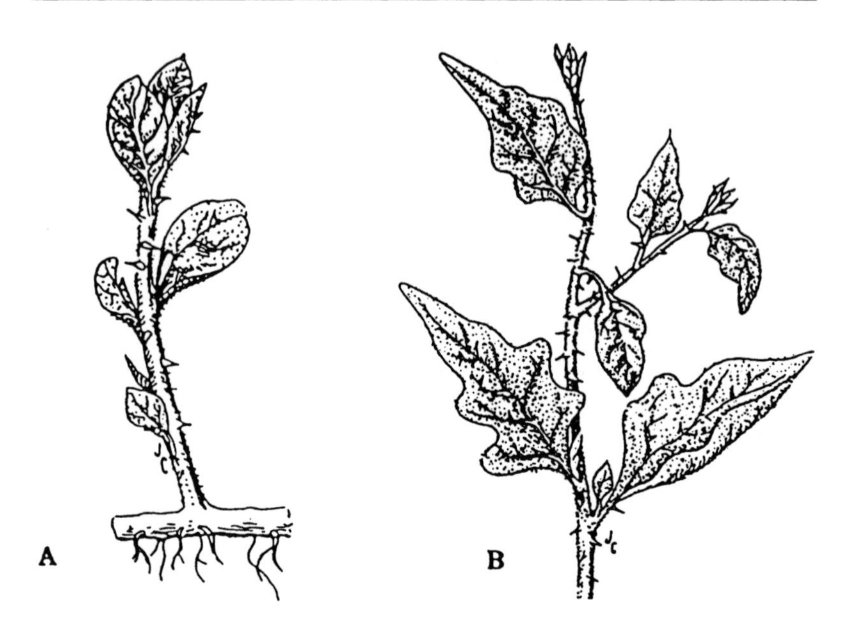

FIGURE 16.5. Horsenettle plant arising from root cutting. A. 2 weeks (×1) and B. 6 weeks (×0.5) after emergence.

Source: Ilnicki, R.D., T.F. Tisdell, S.N. Fertig, and A.H. Furrer, Jr. 1962. Horsenettle. No. 3. In Life History Studies as Related to Weed Control in the Northeast, Bulletin 368, 54 pp. Kingston: Northeast Regional Publication, University of Rhode Island.

2.5 cm) wide, lance-shaped, oblong or linear, with entire or wavy margins. The upper surface of each leaf is green and the lower surface is silvery-green. Both the upper and lower surfaces of the leaves are covered with star-shaped hairs 5 to 15 mm long. The basal leaves of both species are often deciduous.

FLOWERS

The flowers of horsenettle are white, pale blue, or violet to whitish; those of silverleaf nightshade are violet or blue. The flowers of both species have five petals, about 1 inch (2.5 cm) across, the prominent yellow stamens on short filaments united into a conelike shape. The flowers are borne in clusters along the sides of the stem, although at first they appear terminal. They are similar in shape to those of potato and tomato. The plants bloom from April to October in Arizona, and June to September in Nebraska.

SEEDPODS

The seedpods of both species are round berries, pulpy and juicy, hairless and smooth, from 0.67 to 0.75 inch (8 to 20 mm) in diameter. When immature, the fruits are green with dark streaks. The mature berries are yellow, yellowish-brown, or dull orange in color, eventually becoming wrinkled and blackish. A single plant may produce 35 or more berries, and each berry contains 40 to 120 or more seeds.

SEEDS

The seeds of both species are disk-shaped, flat in cross section, 2.0 to 2.4 mm long and 1.5 to 2.5 mm wide, smooth and glossy, and appear minutely granular (pebbly-rough) when viewed through a hand lens. The immature seeds are orange to reddish-brown. The mature seeds are pale to deep yellow or yellowish-orange.

SEEDLINGS

Seedlings of both species can emerge from unbroken berries buried in the upper 1-inch (2.5-cm) soil layer, but seedling emergence occurs more quickly from broken berries. The cotyledons are *epigeous,* pulled aboveground by elongating hypocotyls. The hypocotyl of seedlings of both species is tough, often purple tinged, and densely covered with stiff short hairs. The epigeous cotyledons are small, glossy green on the upper surface and lighter below, and both surfaces are smooth.

POISONOUS

When ingested, the aerial parts of both species are poisonous to cattle, due to the presence of the toxin *solanine* in these parts. The quantity of poison in the aboveground plant parts may be 10 times greater in autumn than during other seasons of the year. The mature berries are the principal poisonous plant part; the immature green fruits are slightly less toxic; the foliage is the least toxic plant part. In the winter months, the mature berries may be preferentially grazed by cattle. Sheep are considerably less susceptible to poisoning than cattle, and goats appear not to be poisoned at all.

ALLELOPATHY

Horsenettle vegetation appears to have allelopathic properties. Leaf tissue had the greatest inhibiting effect on germination of horsenettle seeds, with inhibition occurring with as little as 2 g of leaf tissue mixed with 100 g of soil. The inhibiting effect was only apparent on seed germination, as the buried leaf tissue had no effect on the growth of horsenettle seedlings. The germination inhibitor was leached from the soil with water, and the inhibitory effect on germination decreased with time. This seems to demonstrate that the allelopathic process of inhibition of seed germination, and reversibility of inhibition by leaching, can result in density-dependent regulation of population size in horsenettle.

REFERENCES

Anonymous. 1969. Weeds of Wyoming. Bulletin No. 498. Laramie: University of Wyoming, p. 112.

Bassett, I.J., and D.B. Munro. 1986. The biology of Canadian weeds. No. 78. *Solanum carolinense*. Can. J. Plant Sci. 66: 977–991.

Elmore, D.C. (Chrm). 1985. Weed Identification Guide. Southern Weed Scientific Society. Champaign, IL: SOLCA and 5 SOLEL, p. 5.

Hurst, H.R. 1970. Weeds Common to Row Crops in Arkansas. Fayettville: Agricultural Experimental Station, University of Arkansas, p. 8.

Ilnicki, R.D., T.F. Tisdell, S.N. Fertig, and A.H. Furrer, Jr. 1962. Horsenettle. No. 3. In Life History Studies as Related to Weed Control in the Northeast, Bulletin No. 368. Kingston, RI: Northeast Regional Publication, University of Rhode Island, 54 pp.

Kingsbury, J.M. 1964. Poisonous Plants of the United States and Canada. Englewood Cliffs, NJ: Prentice-Hall, p. 289.

Meister, R.T. (ed). 1997. Weed Control Manual 1997. Willoughby, OH: Meister Publishing.

Parker, K.F. 1972. An Illustrated Guide to Arizona Weeds. Tucson: The University of Arizona Press, pp. 262–263.

Reed, C.F., and R.O. Hughes. 1970. Selected Weeds of the United States. Agriculture Handbook No. 366. Washington, DC: U.S. Department of Agriculture, Superintendent of Documents.

Robbins, W.W., M.K. Bellue, and W.S. Ball. (No date). Weeds of California. Sacramento, CA: State Office of Documents and Publications, pp. 388–390.

Wax, L.M., R.S. Fawcett, and D. Isley. 1990. Weeds of the North Central States. North Central Regional Publication. Bulletin No. 772. Urbana: University of Illinois at Urbana-Champaign, pp. 160.

Westerman, R.B., and D.S. Murray. 1994. Silverleaf nightshade (*Solanum elaeagnifolium*) control in cotton (*Gossypium hirsutum*) with glyphosate. Weed Technol. 8: 720–727.

17

LEAFY SPURGE
(Euphorbia esula)

INTRODUCTION

Leafy spurge (Figure 17.1) is a herbaceous, deep-rooted, long-lived, perennial broadleaf weed. Leafy spurge is not a single species but an aggregation of closely related, perhaps hybridized, taxa. After establishment, leafy spurge tends to form colonies, displacing all other vegetation and establishing single species stands. Colonies of leafy spurge show considerable variation in floral and vegetative morphology. All

FIGURE 17.1. Leafy spurge (*Euphorbia esula* L.). A. Plant habit ×0.5; B. flower cluster ×2.5; C. capsule ×2.5; D. seeds ×6.

Source: Reed, C.F., and R.O. Hughes. 1970. Leafy spurge (*Euphorbia esula* L.). In *Selected Weeds of the United States*. Agriculture Handbook No. 366, p. 249. Washington, DC: U.S. Department of Agriculture, U.S. Government Printing Office.

plant parts contain white, milky latex. The milky latex is a skin irritant that can cause severe dermatitis (irritation, blotching, blistering, swelling) in sensitive humans and grazing animals.

Leafy spurge vegetation is toxic to some animals and unpalatable to most livestock and is generally avoided by grazing animals. Sheep and goats are less affected than other animals by the toxic substance in leafy spurge, and they have been used in leafy spurge control programs.

Leafy spurge is found in grain fields, pastures, ditches, roadsides, sandy banks, wooded and waste areas, and even ungrazed native grassland. It thrives in areas where competing plants are grazed. It often forms dense stands in pastures and rangelands, displacing useful forage plants and restricting cattle grazing. Leafy spurge infestations occur in light sandy to heavy clay soils, but they grow best in coarse-textured soils.

DISTRIBUTION

Leafy spurge, a native to Eurasia, was first reported in North America in 1827 at Newbury, Massachusetts. It was introduced into North America from Europe and Russia. Leafy spurge occurs throughout most of the northern half of the United States, from the Atlantic to the Pacific (Figure 17.2), and in Canada from Nova Scotia to Alberta. It is most abundant in the Northern Great Plains of the United States and the Prairie Provinces of Canada. It is a major problem weed in Montana and Wyoming, and it is becoming more of a problem in Idaho and Oregon. It is not a major pest in the northeastern United States or in adjacent eastern Canada.

RHIZOMES

Leafy spurge plants do not have rhizomes, despite this assertion in some weed identification guides.

PROPAGATION

Leafy spurge propagates by seeds and vegetative regeneration from adventitious root buds.

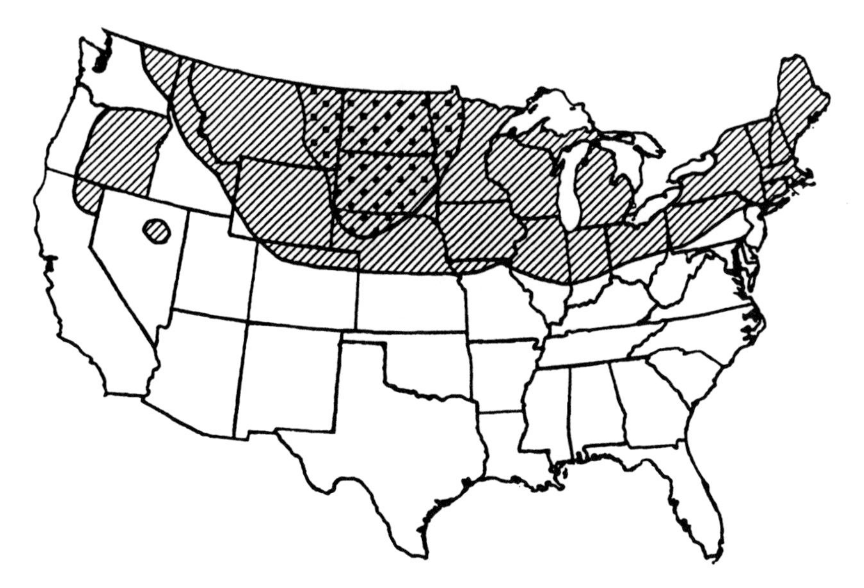

FIGURE 17.2. Distribution of leafy spurge in the United States. Denser cross-hatching denotes infestations of greater economic importance.

Source: Reed, C.F., and R.O. Hughes. 1970. Leafy spurge (*Euphorbia esula* L.). In *Selected Weeds of the United States*. Agriculture Handbook No. 366, p. 248. Washington, DC: U.S. Department of Agriculture, U.S. Government Printing Office.

PERENNATION

The perennial nature of leafy spurge is due to its extensive horizontal and vertical root system and the production of a profusion of new plants from buds on roots, and underground portions (crowns) of vertical stems.

SPREAD

Leafy spurge spreads by seeds and creeping horizontal roots. Seed dispersal is the principal means by which leafy spurge establishes new colonies and spreads to distant areas. Creeping roots account for the spread of leafy spurge within a colony, but the establishment of new colonies by natural root elongation has not been reported.

The spread of leafy spurge in North Dakota is an example of its rapid spread. Leafy spurge was not found in North Dakota until the early 1900s. In the early 1930s, leafy spurge seed was widely spread via infested hay that was distributed in the state during the drought-stricken depression days. By the 1940s, leafy spurge had become a problem weed. In 1962, it had infested an estimated 200,000 A (81,000 ha); in 1973, 423,000 A (171,255 ha); in 1982, 860,000 A (348,178 ha); and in 1987, 1.2 million A (486,000 ha). In 1991, leafy spurge infested about 2.07 million A (840,000 ha) in North Dakota, South Dakota, Montana, and Wyoming.

Plant Description

Leafy spurge is a herbaceous perennial weed with an extensive and persistent vertical and horizontal root system (Figure 17.3). It is one of the first plants to emerge in the spring, normally emerging in March in Iowa and Wisconsin, early April in North Dakota, and the last two weeks of April in Saskatchewan. As a colony develops, its density may reach 1000 plants/yd^2 (0.84 m^2). The plants grow from 8 inches to 3 ft (20 to 90 cm) high, with their main stems branching profusely, resulting in a clumplike appearance. Leafy spurge maintains a heavy growth aboveground throughout the growing season. A milky sap is present in all plant parts, and it will seep from an injured or severed part as a thick, sticky, milky juice.

Roots

Leafy spurge roots are brown with pinkish buds. They are of two basic types; namely, long and short. The long-root types are the vertical and horizontal roots, and they are capable of regeneration. They are woody, have active cambium, exhibit secondary growth, increase in diameter, and persist for more than one growing season. The extensive horizontal and vertical root system of leafy spurge is formed by adventitious roots. Vertical roots may extend to soil depths of 8 to 15 ft (2.5 to 4.5 m).

The short-root types are lateral feeder roots, they do not regenerate nor contribute to the framework of the persistent root system, and their lateral roots usually die at the end of the first growing season. The cambium of the short-root types is not active, a necessity for secondary growth. The network of lateral feeder roots are in the upper 12-inch (30-cm) soil layer.

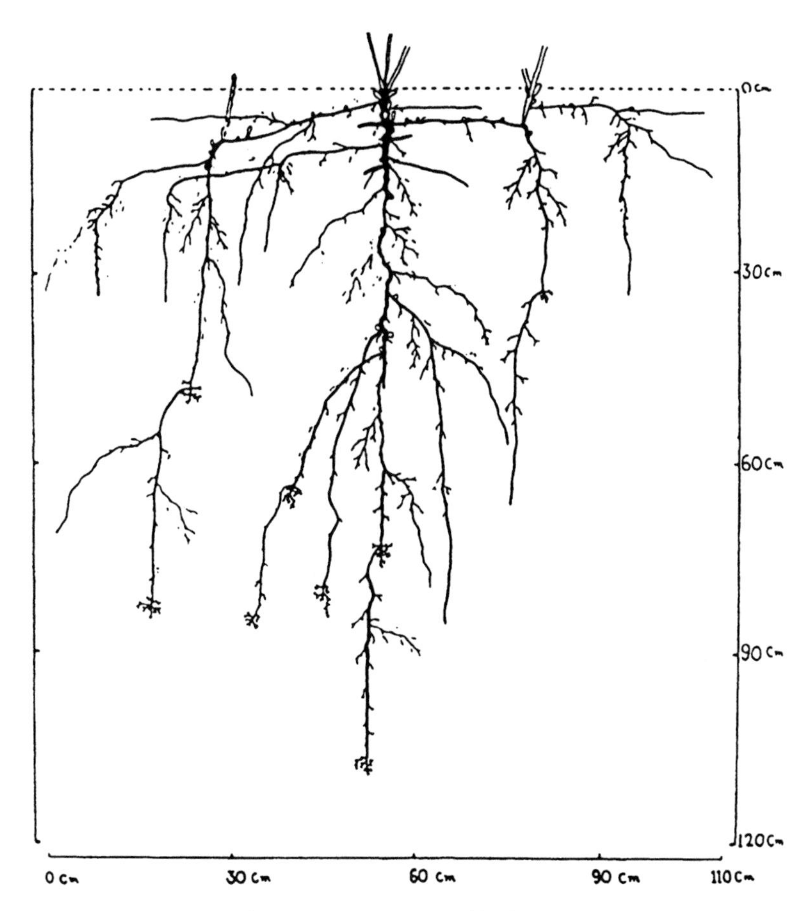

FIGURE 17.3. Diagram of a mature root system of leafy spurge, showing the disposition of horizontal and vertical roots and adventitious aerial shoots.

Source: Raju, M.V.S. 1985. Morphology and anatomy of leafy spurge. In Watson, A.K. (ed.). Leafy Spurge. No. 3. Monograph Series, pp. 26-41. Champaign, IL: Weed Science Society of America. Reprinted by permission.

The upper portion of the plant can be injured or killed by tillage or herbicide treatment, but the root system located below the treatment, or from severed root segments, can develop adventitious buds that can send up new shoots from a depth of 30 cm or more. Root segments 0.5 inch (1.3 cm) or more long and 0.125 inch (3 mm) in diameter are able to produce new shoots within 3 weeks from depths to 12 inches (30 cm) or more. Such small root pieces can withstand exposure to hot sun for 2 to 3 hours before they are killed by desiccation.

Apical Dominance

Although numerous root buds are present, only a few develop into shoots, exhibiting the phenomenon of apical dominance. The early formed vigorous shoots establish dominance over the undeveloped buds. However, early in the season, many buds on underground parts, especially on older stems, develop into aerial shoots without expressing apical dominance.

Shoots

Shoots of leafy spurge are erect, tough, thickly clustered, and 12 to 40 inches (30 cm to 1 m) in height; axillary branches are common. The shoots develop from buds on underground parts of stems, rather than from roots. However, roots produce underground shoots that terminate in crowns at the soil surface. Most aerial flowering shoots arise from adventitious buds on underground parts of older stems, normally located within a few centimeters of the soil surface.

Leafy spurge has a high incidence of shoot emergence all during the growing season. However, shoots emerging in the late summer and fall tend not to grow more than 1 to 5 inches (2.5 to 13 cm) in height. This lack of vegetative growth after emergence is considered an expression of dormancy, because environmental conditions favor growth. However, in the fall, shoot emergence increases after soil temperatures, at a depth of 5 inches (13 cm), have been near freezing for 2 weeks. Dormancy is apparently broken naturally by low soil temperatures.

Shallow cutting of leafy spurge, as by tillage or mowing, may stimulate development of new buds, resulting in a net increase in the number of shoots. For example, leafy spurge regrowth following rototilling averaged 316 shoots/m^2, compared with 134 shoots/m^2 in an undisturbed area. Similar increases have occurred when leafy spurge was treated with herbicides that controlled topgrowth without leaving an effective soil residual to control regrowth.

Leaves

The leaves of leafy spurge are alternate, except for the two opposite leaves at the first two seedling nodes, including the cotyledonary node, and the whorl of leaves subtending the terminal inflorescences. The leaves are highly variable in shape, ranging from broadly linear-lanceolate to ovate. They are narrow, 1 to 4 inches (2.5 to 10 cm) long, without petioles or hairs, and pale green or bluish-green in color. In the fall, as the plants mature, the stems and leaves often change in color from bluish-green to a reddish-brown, red, or yellow. Most leaves of leafy spurge fall from the plants prior to a killing frost.

FLOWERS

Leafy spurge produces showy yellow bracts during late May or early June; these may be mistaken for flower petals. When photographed using conventional color or color infrared film, the conspicuous yellow bracts produce a distinct yellow-green or pink image, respectively. On conventional color video imagery, the yellow bracts produce a golden-yellow image. Use of this color imagery in aerial photography permits locating leafy spurge infestations and monitoring the spread or decline of leafy spurge populations.

Leafy spurge plants from seedlings and asexual buds generally do not produce flowers until the second year. The flowers are small and borne in a flowerlike inflorescence called a *cyathium*, a special flower-type characteristic of the genus *Euphorbia*. The inconspicuous greenish-yellow blossoms borne in umbrella-like clusters are more like dense tufts of small leaves than ordinary flowers. Leafy spurge pollen is sticky, enhancing cross-pollination by insects.

In an established stand of leafy spurge, flowering occurs throughout the growing season. Early flowering shoots of leafy spurge appear in May, and their terminal inflorescences become obvious in June. Their fruits (capsules) mature by about the middle of July. Late flowering shoots will emerge in late June to early July and produce terminal inflorescences in August and mature fruits in September/October. Both the early flowering and late flowering shoots also produce axillary branches that terminate in inflorescences.

Flower production and seed development are continuous from late May or early June through late July into early August, depending on location and climatic condition. Seed development and maturation continue for approximately 30 days after appearance of the last flower. The plants often cease to grow and bloom during the hottest and driest parts of the year, usually July and August.

Seeds

Seeds of leafy spurge are oblong, ranging in size from 2.0 to 2.5 mm long and 1.3 to 1.5 mm wide, smooth, light-gray to yellow-brown, with a characteristic yellow (or white) carbuncle at one end. The seeds are contained in a three-cell capsule. The number of seeds per capsule varies; nearly 50% produce only one mature seed, about 35% produce two seeds, and only 15% to 20% produce three seeds. Individual flower stems normally produce from 10 to 50 capsules, with an average of 220 seeds per stem.

In Saskatchewan, Canada, the average seed production in the center of leafy spurge patches was 2500 seeds/m^2, ranging from 790 to 8020 seeds/m^2. The capsules begin to dehisce about mid-July and continue until late fall. The peak period for seed maturity is from mid-to-late July in most locations.

The seeds are forcibly ejected from capsules up to 13 ft (4 m) or more from the parent plant. Seed dormancy permits germination for a period up to 5 years following maturity. Early spring (about May) is the most favorable time for seedling emergence, even when temperatures are near freezing. However, seeds germinate and seedlings emerge throughout the entire growing season if moisture is adequate.

Leafy spurge seed can float and germinate on water. In the field, nondormant leafy spurge seeds can germinate in the spring as soon as temperature and moisture conditions are favorable. The seeds imbibe water quickly, which contributes to their rapid germination.

Seeds produced by leafy spurge vary with age in color and viability. After fertilization, and as the seeds increase in age, their color changes from yellow to yellow with brown tips, to brown ends with a narrow yellow band between (the brown moves in from both ends), to brown with an orange band, to reddish-brown, and then followed in order by brown, gray-brown, gray, and finally, mottled. Seeds with coats having any one of the last four colors are viable; the others are not. Brown seeds appear 10 to 13 days after fertilization, depending on seasonal conditions. Harvested in June and in August, the germination of brown seeds is only 25%, whereas that of gray and mottled seeds is 88% and 81%, respectively. New inflorescences are being produced continually during the season, which accounts for seeds of different ages and colors on the same plants.

SEEDLINGS

Leafy spurge seedlings can emerge through 0.5 to 2 inches (1.3 to 5 cm) of soil. Seedling emergence from the upper 1 cm of soil is generally low, perhaps due to low moisture or high temperatures near the soil surface. The seedlings develop rapidly after emergence. The cotyledons are pulled aboveground as the arched hypocotyl becomes erect; such cotyledons are called *epigeous*. The roots can be 24 inches (61 cm) long and the stem 5 inches (13 cm) tall within 2 months after the cotyledons expand. The cotyledons and first two leaves are opposite, but the remaining leaves are distinctly alternate. Young seedlings appear purplish-pink due to anthocyanin pigments in the hypocotyl, eventually turning green. Vegetative buds develop on each seedling just above the soil surface 10 to 12 days after emergence or when six leaves are present.

The seedlings are capable of vegetative reproduction 7 to 10 days after emergence. Ordinarily, adventitious buds form at the hypocotyl/root transition zone of the seedlings 2 to 3 days after the seedlings emerge aboveground. Additional adventitious buds develop on the primary root early in the development of the seedlings, and as the seedlings age, the number of buds on the primary root increases. As the season progresses, the seedling shoots dry up and appear dead, but their underground parts persist. With little or no competition, seedling shoots may survive until the end of the first growing season. More commonly, they are soon replaced by adventitious shoots which, with their seedling root system, persist in the field. Adventitious shoots increase in number from early in the growing season until about the middle of June. By early July, there is a decrease in the number of vegetative shoots and a concurrent increase in the number of flowering shoots, indicating the transformation of vegetative shoot apices to floral apices. The original shoots never mature into flowering stems.

REFERENCES

Alley, H.P., and C.G. Messersmith. 1985. Chemical control of leafy spurge. In Watson, A.K. (ed), Leafy Spurge. Weed Science Society Monograph Series, No. 3. Champaign, IL: Weed Science Society of America, pp. 65–78.

Derscheid, L.A., G.A. Wicks, and W.H. Wallace. 1963. Cropping, cultivation, and herbicides to eliminate leafy spurge and prevent reinfestations. Weeds 11: 105–111.

Derscheid, L.A., J. Wrage, and W.E. Arnold. 1985. Cultural control of leafy spurge. In Watson, A.K. (ed), Leafy Spurge. Weed Science Society Monograph Series, No. 3. Champaign, IL: Weed Science Society of America, pp. 57–64.

Dunn, P.H. 1985. Origins of leafy spurge in North America. In Watson, A.K. (ed.), Leafy Spurge. Weed Science Society Monograph Series, No. 3. Champaign, IL: Weed Science Society of America, pp. 7–13.

Everitt, J.H., G.L. Anderson, D.E. Escobar, M.R. Davis, N.R. Spencer, and R.J. Andrascik. 1995. Use of remote sensing for detecting and mapping leafy spurge (*Euphorbia ensula*). Weed Technol. 9: 599–609.

Gylling, S.R., and W.E. Arnold. 1985. Efficacy and economics of leafy spurge (*Euphorbia ensula*) in pasture. Weed Sci. 33: 381–385.

Harris, P., P.H. Dunn, D. Schroeder, and R. Vonmoos. 1985. Biological control of leafy spurge in North America. In Watson, A.K., (ed.), Leafy Spurge. Weed Science Society Monograph Series, No. 3. Champaign, IL: Weed Science Society of America, pp. 72–92.

Lajeunesse, S., R. Sheley, R. Lym, D. Cooksey, C. Duncan, J. Lacey, N. Rees, and M. Ferrell. 1995. Leafy spurge: Biology, Ecology, and Management. Bulletin W-1088, EB 134. Bozeman: Montana State University Extension Service, 25 pp.

Lym, R.G., C.G. Messersmith, and R. Zollinger. 1993. Leafy Spurge: Identification and Control. Bulletin W-765 (Revised). Fargo: North Dakota State University Extension Service, 7 pp.

Lym, R.G., and C.G. Messersmith. 1985. Leafy spurge control with herbicides in North Dakota: 20-year survey. J. Range Management 38: 149–154.

Messersmith, C.G., R.G. Lym, and D.S. Galitz. 1985. Biology of leafy spurge. In Watson, A.K. (ed), Leafy Spurge. Weed Science Society Monograph Series, No. 3. Champaign, IL: Weed Science Society of America, pp. 42–56.

Meyers, G.A., C.A. Beasly, and L.A. Derscheid. 1964. Anatomical studies of *Euphorbia ensula* L. Weeds 12: 291–295.

Mitich, L.W. 1966. Leafy spurge: A problem weed controlled by Tordon. Down To Earth 21(4): 11–13.

Mitich, L.W. 1967. Control of leafy spurge, field bindweed, and western sowberry with Tordon herbicide. Down To Earth 23(3): 8–11.

Monson, W.G., and F.S. Davis. 1964. Dormancy in western ironweed and leafy spurge. Weeds 12: 238–239.

Raju, M.V.S. 1985. Morphology and anatomy of leafy spurge. In Watson, A.K.(ed), Leafy Spurge. Weed Science Society Monograph Series, No. 3. Weed Science Society America, Champaign, IL: pp. 26–41.

Reed, C.F., and R.O. Hughes. 1970. In Selected Weeds of the United States. Agricultural Handbook No. 336. Washington, DC: United States Department of Agriculture, Superintendent of Documents, pp. 248–249.

Rees, N.E., P.C. Quimby, Jr., G.L. Piper, E.M. Coombs, C.E. Turner, N.R. Spenser, and L.V. Knutson (eds). 1995. Leafy spurge. In Biological Control of Weeds in the West. Bozeman: Western Society of Weed Science, Montana State University.

Sedivec, K., T. Hanson, and C. Heiser. 1995. Controlling leafy spurge using goats and sheep. Bulletin R-1093. Fargo: Extension Service, North Dakota State University.
Watson, A.K. 1985. Introduction: The leafy spurge problem. In Watson, A.K. (ed.), Leafy Spurge. Weed Science Society. Monograph Series, No. 3. Champaign, IL: Weed Science Society of America, pp. 1–6.
Watson, A.K. 1985. Integrated management of leafy spurge. In Watson, A.K. (ed), Leafy Spurge. Weed Science Society Monograph Series, No. 3. Champaign, IL: Weed Science Society of America, pp. 93–104.
Wicks, G.A., and L.A. Derscheid. 1964. Leafy spurge seed maturation. Weeds 12: 175–176.

18

Red Sorrel
(Rumex acetosella)

Introduction

Red sorrel (Figure 18.1), also called *sheep sorrel,* is a creeping, herbaceous, perennial broadleaf weed. In the United States, red sorrel is listed as a restricted noxious weed in the seed laws of 18 states. Its seeds are difficult or impossible to remove completely from some kinds of grass and clover seeds, and such crop seeds are likely to be discounted in price when sold. Red sorrel is most troublesome in gardens,

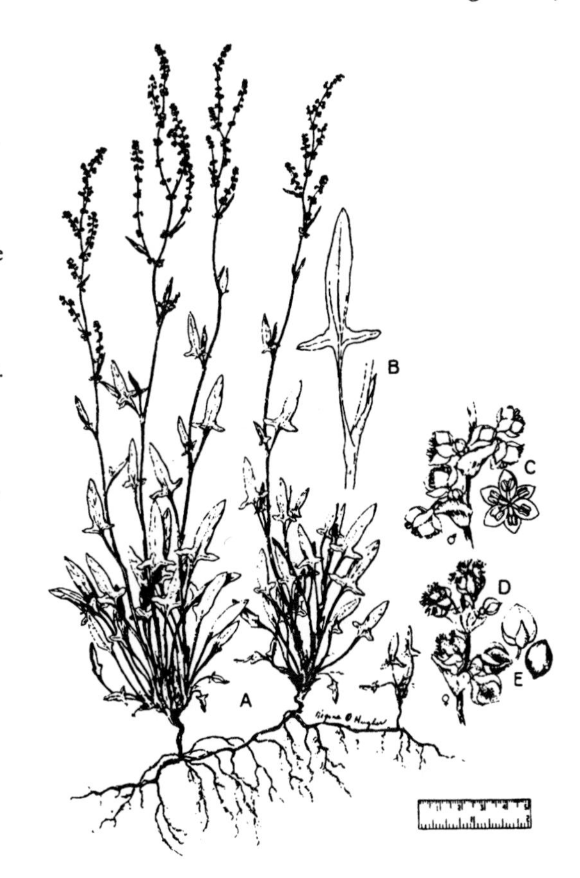

Figure 18.1. Red sorrel (*Rumex acetosella* L.). A. Plant habit ×0.5; B. leaf detail ×1.5; C. staminate (male) flowers ×7.5; D. pistillate (female) flowers ×7.5; E. achenes (seeds), in and out of calyx ×10.

Source: Reed, C.F., and R.O. Hughes. 1970. Red sorrel (*Rumex acetosella* L.). In *Selected Weeds of the United States.* Agriculture Handbook No. 366, p. 129. Washington, DC: U.S. Department of Agriculture, U.S. Printing Office.

lawns, strawberry plantings, grass or clover seed fields, and pastures. Its growth is favored by the absence of plant competition, and conditions encountered in unproductive acid soils of low fertility and poor drainage. Alkaline soils are unfavorable to survival of red sorrel seedlings.

DISTRIBUTION

Red sorrel is naturalized from Eurasia. It is distributed throughout Europe. It is a common weed found throughout the United States (Figure 18.2) and southern Canada.

RHIZOMES

Red sorrel plants do not have rhizomes, despite this assertion in some weed identification guides.

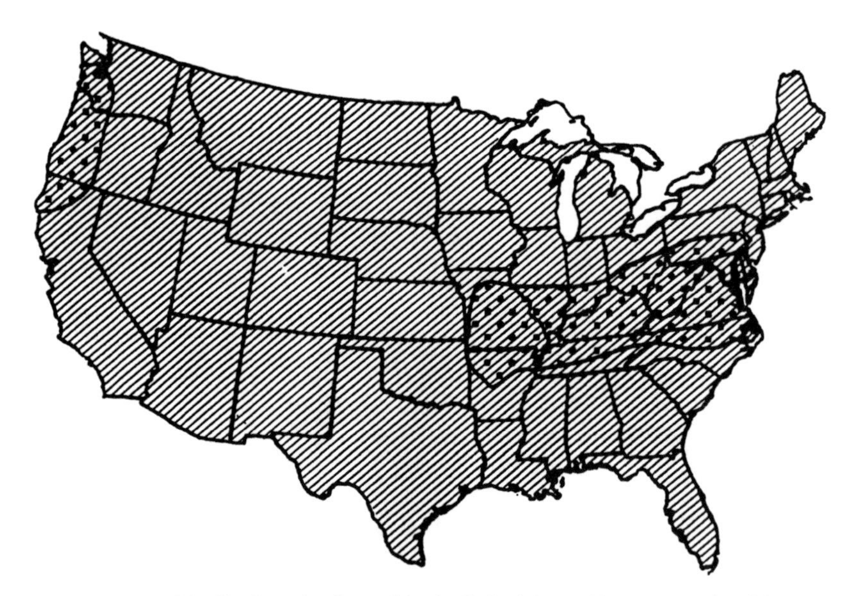

FIGURE 18.2. Distribution of red sorrel in the United States. Denser cross-hatching denotes infestations of greater economic importance.

Source: Reed, C.F., and R.O. Hughes. 1970. Red sorrel (*Rumex acetosella* L.). In *Selected Weeds of the United States*. Agriculture Handbook No. 366, p. 128. Washington, DC: U.S. Department of Agriculture, U.S. Government Printing Office.

PROPAGATION

Red sorrel reproduces by seeds and by adventitious buds on roots (Korsmo, 1954) forming an extensive, but shallow, horizontal root system.

SPREAD

Red sorrel spreads by seeds and shallow creeping roots.

PERENNATION

Adventitious buds on the overwintering root system are responsible for the perennial nature of red sorrel.

PLANT DESCRIPTION

Red sorrel is a herbaceous, perennial broadleaf weed with an extensive, but rather shallow, feeder-root and widely branched root system. The roots are yellowish and creeping. During development of the root system, some roots spread out shallowly, while others go deeper, turning into deep vertical roots. The horizontal roots are often mistaken for rhizomes. The plants are highly variable in size and color. The stems are slender, erect, 6 to 24 inches (15 to 60 cm) tall, somewhat woody at the base, and branched at the top. New shoots arise from both roots and crowns, form rosettes of leaves, and several stems may arise from the same crown. Red sorrel plants contain oxalic acid, giving the stems and leaves an acrid taste; the sap causes dermatitis in some animals.

Seedlings form a rosette of basal leaves in early growth and leaf shapes change with age. Young leaves are egg-shaped, the margins are entire, and the bases taper into the petiole; both the upper and lower surfaces are covered with waxy granules. The characteristic basal lobes develop on the third, fourth, and fifth leaves.

The cotyledons are *epigeous,* pulled aboveground by elongating hypocotyls. They are oblong, dull green, 10 mm long, and the petioles are flattened on the upper surface and united basally.

The leaves are alternate, simple and entire, thick, smooth, and lanceolate (arrow-shaped), with slender petioles having a papery sheath at the point of attachment to the stem. The leaves are 1 to 4 inches (2.5 to 10 cm) long, including the petiole. The lower leaf blades have two conspicuous basal lobes; the upper leaves are more slender and sometimes without the basal lobes. The leaves are acid to the taste.

The plants are dioecious, with male and female flowers on different plants. The male flowers are orange-yellow in color and the female flowers are red-orange. The inflorescence is a slender raceme, and the flowers are borne in terminal, branched clusters (panicles). The inflorescences terminate the stems. The flowers have no corolla (petals). The fruits are small, three-angled, enclosed in three reddish, persistent flower parts. The triangular seeds are a shiny reddish-brown to golden brown and about 1.5 mm long. The hulls are reddish-brown, rough, often adhering to the seed. Red sorrel flowers from June to October.

REFERENCES

Freeman, J.F. 1972. Red sorrel as a weed. Weeds Today 3(4): 68.

Gaines, X.M., D.G. Swan, and H.C. Keller. 1972. Red sorrel (*Rumex acetosella*). In Weeds of Eastern Washington and Adjacent Areas. Davenport, WA: Camp-Na-Bor-Lee Association, Inc., pp. 80-81.

Korsmo, E. 1954. *Rumex acetosella*. In Anatomy of Weeds. Forlag, Norway: Grondahl and Sons, pp. 90-93.

Murphy, T.R. 1993. Red sorrel (*Rumex acetosella*). In Weeds of Southern Turf Grasses. Athens, GA: Agricultural Business Office, The University of Georgia, p. 165.

Nelson, E.W., and L. Robison. 1968 (revised). Red sorrel (*Rumex acetosella* L.). In Nebraska Weeds. Bulletin No. 101-R. Lincoln: Nebraska Department of Agriculture, p. 17-3.

Parker, K.F. 1972. Red sorrel (*Rumex acetosella*). In An Illustrated Guide to Arizona Weeds. Tucson: The University of Arizona Press, pp. 86-87.

Reed, C.F., and R.O. Hughes. 1970. Red sorrel (*Rumex acetosella*). In Selected Weeds of the United States. Agricultuiral Handbook No. 366. Washington, DC: U.S. Department of Agriculture, Superintendent of Documents, pp. 128-129.

Wax, L.M., R.S. Fawcett, and D. Isely. 1990. Red sorrel (*Rumex acetosella*). In Weeds of the North Central States. North Central Regional Research Publication No. 281, Bulletin 772. Urbana: College of Agriculture, University of Illinois at Urbana-Champaign, p. 53.

Whitson, T.D. (ed). 1991. Red sorrel (*Rumex acetosella*). In Weeds of the West. Laramie: The Western Society of Weed Science, Bulletin Room, University of Wyoming, p. 513.

Part Seven

Perennial Broadleaved Weeds Reproducing from Buds on Creeping, Horizontal Rhizomes

19

Stinging Nettle
(Urtica dioica)

Introduction

Stinging nettle (Figure 19.1) is a herbaceous, creeping, perennial broadleaf weed. It is native to North America. It grows in large colonies, rarely as a single plant with one or a few stems. A conservative estimate of the lifespan of undisturbed stinging nettle clones is 50 years or more. Stinging nettle infestations occur in cultivated row crops, gardens, orchards, nurseries, old pastures, farm yards, along roadsides, drainage ditches, irrigation canals, and stream banks.

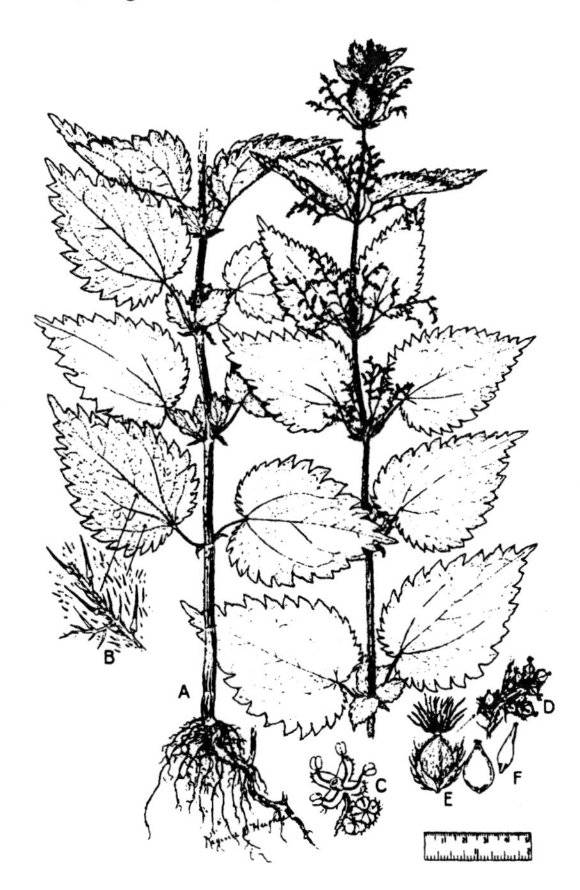

FIGURE 19.1. Stinging nettle (*Urtica dioica* L.). A. Plant habit ×0.5; B. stinging nettle hairs enlarged; C. flower, closed and open ×7.5; D. fruiting spike ×1.5; E. fruit ×6.5; F. achenes, face and edge views ×7.5.

Source: Reed, C.F., and R.O. Hughes. 1970. Stinging nettle (*Urtica dioica* L.). In *Selected Weeds of the United States*. Agriculture Handbook No. 366, p. 111. Washington, DC: U.S. Department of Agriculture, U.S. Government Printing Office.

Stinging nettle is primarily a nuisance weed because of its stinging hairs. The hairs on the stem, leaves, and flowers produce a painful sting to human flesh, followed by an immediate small reddish swelling, prolonged itching, and numbness of the sensitized area. The painful sting normally lasts a few minutes, but if stung repeatedly, the pain often intensifies and can last for several days. The allergenic reaction of humans to stinging nettle varies; some appear completely immune while others are highly sensitive.

Pollen of stinging nettle, shed in large amounts, is an important contributor to summer hay fever, with many people more allergic to its pollen than to that of common ragweed (*Ambrosia artemisiifolia*). In Minnesota, a 32-year analysis of atmospheric pollen ranked *Urtica* spp. 10th in abundance among 50 plant genera.

Stinging nettle plants prefer soils rich in nutrients. The plants will not persist in saline areas or in habitats of low soil fertility, especially those with low phosphate levels. The permeability of stinging nettle root cells to phosphorus is reduced by calcium. Populations of stinging nettle will grow in soils ranging in pH from 5.6 to 7.6.

DISTRIBUTION

Stinging nettle is found throughout much of the United States, except for southern Georgia, most of Florida, and states to the west of a diagonal line from northwestern Washington through most of Texas (Figure 19.2). Stinging nettle is a native plant to Oregon, and it occurs along stream courses in arid regions and on moist shaded lowlands or mountain slopes; its presence in Oregon is not adequately illustrated in Figure 19.2. Local infestations of stinging nettle may be found in Arizona, California, and eastern Utah.

In Canada, stinging nettle occurs in all provinces, as well as the Yukon Territory, and Mackenzie District, N.W.T. Although widely distributed, it is abundant mainly in lowland situations, and it is often associated with human habitation. Infestations often occur in rich soils suited for gardens, nurseries, and orchards. A large population of American stinging nettle extends for 9 mi (15 km) along the highway embankments between Kenora, Ontario, and Winnipeg, Manitoba.

Stinging nettle is a pestiferous weed throughout Great Britain and

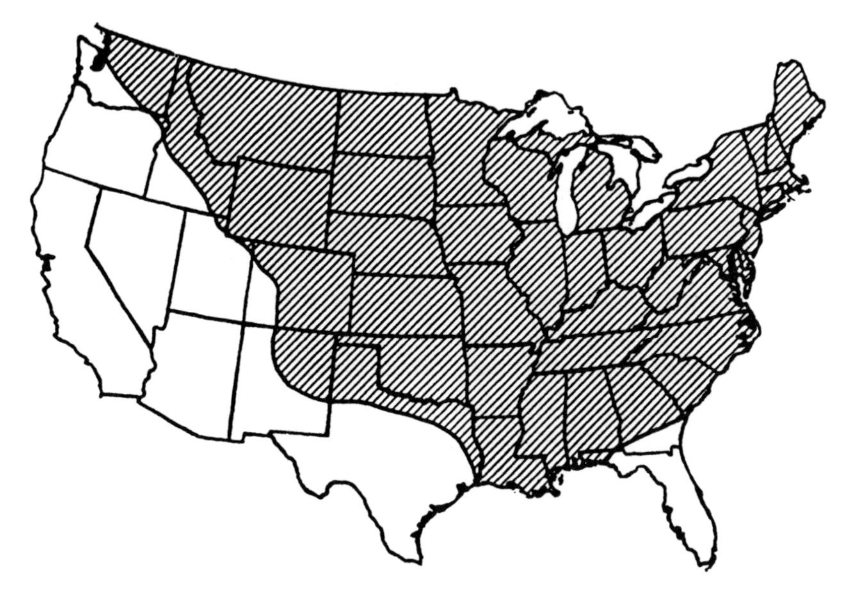

FIGURE 19.2. Distribution of stinging nettle in the United States.

Source: Reed, C.F., and R.O. Hughes. 1970. Stinging nettle (*Urtica dioica* L.). In *Selected Weeds of the United States*. Agriculture Handbook No. 366, p. 110. Washington, DC: U.S. Department of Agriculture, U.S. Government Printing Office.

Ireland. The author got his introduction to the plant by brushing against a leaf in a patch alongside an urban pathway in Pitlochry, Scotland. Thereafter, identification was immediate, based on leaf shape, light-green color, and erect plants growing in clusters.

PROPAGATION

Stinging nettle reproduces by seeds and adventitious buds on creeping rhizomes.

SPREAD

Stinging nettle spreads by seeds and creeping rhizomes.

PERENNATION

Rhizomes are responsible for the perennial nature of stinging nettle.

Rhizomes

The vigorous, creeping rhizomes of an established stinging nettle plant radiate from the parent plants to form dense stands of shoots that virtually exclude other species from the immediate area. Maximum rhizome development occurs in the late spring and early summer, prior to flowering and subsequent seed set. When rhizome segments are scattered in a field by disturbances such as cultivation, new plants and, subsequently, colonies arise from these scattered rhizome segments.

A seedling will form a perennating rhizome in the first growing season. A portion of a rhizome planted in late summer can produce a clone 8 ft (2.5 m) in diameter by the following year.

Two Subspecies of Stinging Nettle

Two subspecies, one native and the other introduced, of stinging nettle have been identified in Canada. The native subspecies is *Urtica dioica* L. ssp. *gracilis*, and it is called "common American stinging nettle" (also, tall nettle). The introduced subspecies is *Urtica dioica* L. ssp. *dioica*, and it is called "European stinging nettle." American stinging nettle is widespread throughout Canada, while European stinging nettle is confined to a few disturbed habitats in eastern Canada. In this chapter, reference to stinging nettle refers to common American stinging nettle. The two subspecies can be distinguished one from the other by the following:

Plants of common American stinging nettle are monoecious (bearing staminate and pistillate flowers on the same plant). The stems are erect, glabrous, and rigid. The stinging hairs are usually only on the lower leaf blades.

Plants of European stinging nettle are predominantly dioecious (male and female flowers borne on different plants). The stems are weak and the plants have a sprawling, branching habit. The stems and leaf blades are usually covered with stinging hairs on all surfaces.

In the United States, plants of stinging nettle have apparently not been separated into two subspecies, as in Canada. Some weed identification guides do point out that male and female flowers are borne separately on the same plant (monoecious). However, other guides incorrectly state that plants of stinging nettle are dioecious. This mistake may be due to the fact that male flowers are produced in greater abundance during June and July and female flowers appear mostly later in the growing season.

STINGING HAIRS

The hairs (Figure 19.3) of stinging nettle have been described as being similar to a hypodermic needle. The hairs are hollow, and at their base is a bulb resting in a cuplike socket, and filled with the stinging agent. The bulb and socket appear similar to an egg resting in an old-fashioned egg cup. The tip of the hair is brittle, with sealed tips. The slightest pressure of human skin against a hair will cause it to penetrate the skin, simultaneously breaking off the hair's brittle tip, and forcing the bulb down against the cup-shaped socket, resulting in the stinging agent being injected into the skin through the tip of the hair. This action can be envisioned as similar to the plunger of a hypodermic syringe forcing liquid through its hollow needle.

The agent causing the sting of stinging nettle was thought to be formic acid, which also causes the sting of an ant bite. However, it now

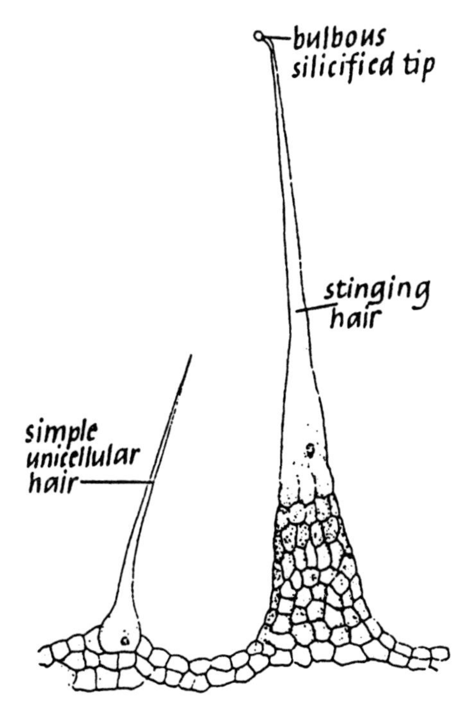

FIGURE 19.3. Diagram of a stinging hair from a leaf of stinging nettle.

Source: Brook, A.J. 1964. *The Living Plant*. Chicago: Aldine Publishing Company, p. 118.

appears to be due to acetylcholine in combination with hydroxytriptamine and an unnamed histamine.

DISTINGUISHING CHARACTERISTICS

The stems of stinging nettle are erect, sometimes with a few ascending branches, four-angled (squarish), 3 to 6 ft (1 to 2 m) or more tall, glabrous except for stinging hairs 0.75 to 2 mm long.

The leaves are opposite, simple, broad at the base, coarsely toothed, usually twice as wide as the length of the petiole, 1.5 to 6 inches (4 to 15 cm) long, with stinging hairs. The leaves have slender petioles with small leaflike parts (stipules) where the petiole attaches to the stem. The stipules are linear to lanceolate, 5 to 15 mm long, green to pale brown, minutely pubescent.

The cotyledons are *epigeous*, pulled aboveground by elongating hypocotyls. They are oval, notched at the apex, 1 mm wide and 1.5 to 4 mm long, with a few short hairs on the upper surface.

The inflorescences are branched, many flowered, loose to dense, panicled spikes. The flowers are imperfect (have either stamens or pistils but not both). The two types of flowers are borne on the same plant and shoots. Both types of flowers are small, greenish, and borne at the nodes on four, slender, branching panicles clustered in the upper leaf axils between the petiole and stem. Staminate flowers have four perianth segments (sepals and petals) and four stamens. Pistillate flowers have four perianth parts and a one-celled ovary. The seed (achene) is 1 to 1.5 mm long, flattened, egg-shaped, minutely glandular, and yellow to grayish-tan. Plants flower from May to September. The flowers are normally wind pollinated.

Interestingly, as the staminate flowers mature, the anthers dehisce (split), forcing the perianth parts to open. The filaments (slender stalks supporting the anthers), which are bent springlike, are released, and the pollen grains are thrown into the air from the open anthers.

SEEDS

Stinging nettle produces one seed per flower. Each cluster of four panicles produces approximately 1000 seeds. Plants growing in the shade have loose, drooping, panicles producing 500 to 5000 seeds per shoot. Plants growing in sunlight have compact, stiff panicles that produce

10,000 to 20,000 seeds per shoot. The seeds normally remain on the plant until frost, when they drop to the ground. Mature seeds collected directly from a plant will germinate in 5 to 10 days.

REFERENCES

Bassett, I.J., C.W. Crompton, and D.W. Woodland. 1977. The biology of Canadian weeds. 21. *Urtica dioica* L. Can. J. Plant Sci. 57: 491–498.
Brooks, A.J. 1964. The Living Plant. Chicago: Aldine Publishing Company, p. 118.
Elliott, C. 1997. Rash encounters with stinging nettles. Horticulture 94(1): 30, 32.
Gaines, X.M., and D.G. Swan. 1972. Weeds of Eastern Washington and Adjacent Areas. Davenport, WA: Camp-Na-Bor-Lee Association, p. 72.
Hawkes, R.B., T.D. Whitson, and L.J. Dennis. 1985. A Guide to Selected Weeds of Oregon. Corvallis: Oregon Department of Agriculture, Oregon State University, p. 18.
Korsmo, E. 1954. *Urtica dioica.* In Anatomy of Weeds, Forlag, Norway: Grondahl and Sons, pp. 72–75.
Poling, J. 1971. Leaves: Their Amazing Lives and Strange Behavior. New York: Holt, Rinehart, and Winston, pp. 53–54.
Reed, C.F., and R.O. Hughes. 1970. Stinging nettle (*Urtica dioica* L.). In Selected Weeds of the United States. Agriculture Handbook No. 366. Washington, DC: U.S. Department of Agriculture, Superintendent of Documents, pp. 110–111.
Thornton, B.J., and H.D. Harrington. (No date). Weeds of Colorado. Bulletin 514-S. Fort Collins: Colorado State University, p. 53.
Wax, L.M., R.S. Fawcett, and D. Isely. 1990. Weeds of the North Central States. Bulletin 772. Urbana: University of Illinois at Urbana-Champaign, p. 48.
Whitson, T.D. 1991. Weeds of the West. Laramie: The Western Society of Weed Science and Cooperative Extension of the Western States, University of Wyoming, p. 589.

20

Western Ironweed
(Vernonia baldwinii)

Introduction

Western ironweed (Figure 20.1) is an erect, herbaceous, perennial broadleaf weed. It is a weed of prairies, pastures, woods, roadsides, and waste areas of the central United States. Under conditions of low soil fertility and overgrazing, western ironweed often becomes the dominant species in an infested area.

Distribution

Western ironweed is native to the United States. It occurs in the central United States, extending west as far as Colorado and New Mexico and east as far as West Virginia and Virginia (Figure 20.2).

Distinguishing Characteristics

Western ironweed characteristically grows in a cluster of velvety stems 3 to 6 ft (1 to 2 m) tall, varying in number from a few to more than 200 stems. The plants have a vigorous fleshy root system that penetrates the soil to depths of 8 ft (2.4 m) or more. These roots arise adventitiously from the rhizomes, and they are almost devoid of secondary roots. Secondary roots do not produce aerial shoots. The leaves are alternate, 4 to 8 inches (10 to 20 cm) long, two to three times longer than broad, widest in the middle and tapering to a pointed apex, rough-bristly above and velvety below. The inflorescence is corymblike to paniclelike with 18 to 34 flowers per head. The seeds (achenes) are about 3mm long. The plants flower from July to September. The aerial shoots of western ironweed are killed each year by freezing temperatures.

Propagation

Western ironweed propagates by seeds and axillary buds borne on long-lived, but short (3 to 12 inches; 7.5 to 30 cm), woody rhizomes.

Figure 20.1. Western ironweed (*Vernonia baldwinii* Torr.). A. Plant habit ×0.5; B. tomentose underside of leaf, enlarged; C. flower head ×3; D. single flower ×6; E. achenes ×6.

Source: Reed, C.F., and R.O. Hughes. 1970. Western ironweed (*Vernonia baldwinii* Torr.). In *Selected Weeds of the United States*. Agriculture Handbook No. 366, p. 443. Washington, DC: U.S. Department of Agriculture, U.S. Government Printing Office.

Spread

Western ironweed spreads primarily by seeds. Its rhizomes are short and do not creep extensively.

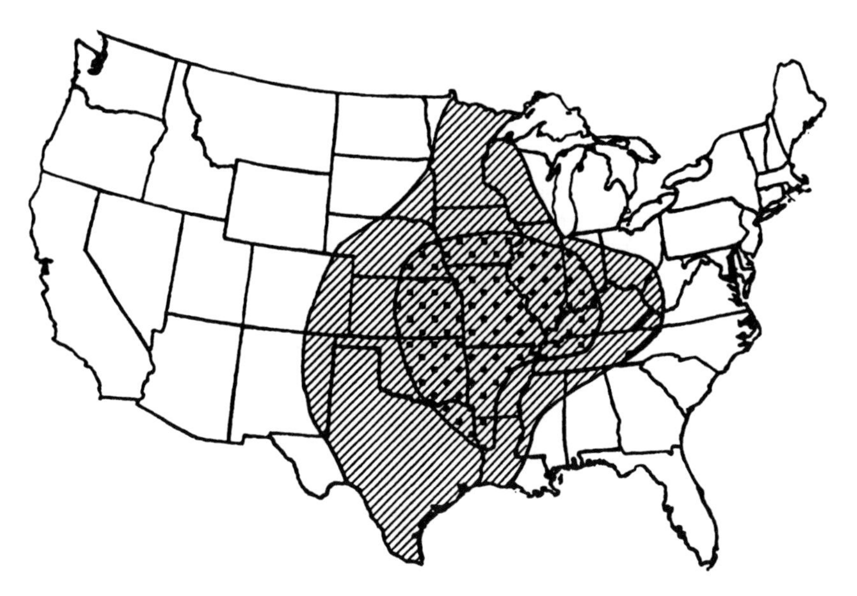

FIGURE 20.2. Distribution of western ironweed in the United States. Denser cross-hatching denotes infestations of greater economic importance.

Source: Illustrations by Lucretia B. Hamilton from *An Illustrated Guide to Arizona Weeds* by Kittie Parker. Copyright ©1972 The Arizona Board of Regents. Reprinted by permission of the Univerity of Arizona Press.

PERENNATION

Rhizomes are responsible for the perennial nature of western ironweed, overwintering and reproducing from axillary buds the next season.

RHIZOMES

Western ironweed plants have short rhizomes. The rhizomes grow 2 to 3 inches (5 to 7.5 cm) in length annually. These annual growth increments are visible as stem scar remnants which allow dating of the rhizome segments. A number of lateral buds, capable of developing into additional rhizomes or shoots, are concentrated along the rhizome (Figure 20.3). Usually, only one aerial shoot per rhizome segment is produced each year. However, the shoots may form such tight clusters as to appear to arise from the same rhizome segment, but if the clusters are washed for inspection one will note that only one to a few

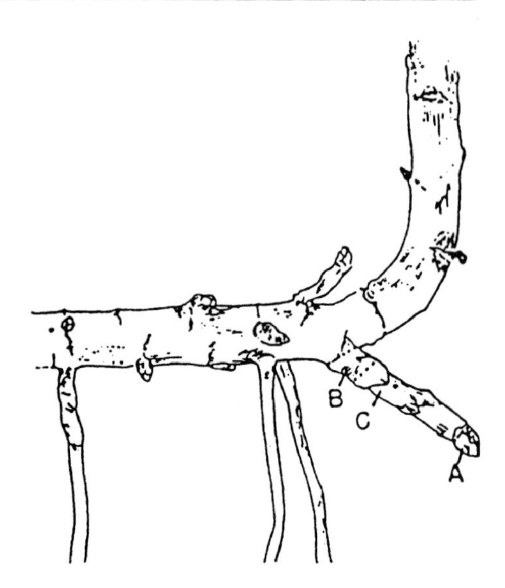

FIGURE 20.3. Diagram of a portion of a mature rhizome showing an aerial shoot and the development of a dominant young rhizome on a western ironweed plant. The letters *A, B,* and C designate apical, basal, and central portions of the young rhizome, respectively. Note the extraneous buds on the stem and older rhizome.

Source: Scifres, C.J., and M.K. McCarty. 1966. Effect of several factors on the expression of dormancy in western ironweed. *Weed Science* 14: 62-69.

shoots arise from a single rhizome. The rhizomes grow in the upper 2- to 3-inch (5- to 7.5-cm) soil layer.

SEEDLINGS

Western ironweed seedlings normally emerge from the soil in late April to early May. A 6-day-old seedling is diagramed in Figure 20.4. The development of seedlings during the first 8 weeks after germination is illustrated in Figure 20.5. They assume their perennial character early in the seedling year.

The seedling's hypocotyl and taproot are separated by a conspicuous corticular collar, a region called the hypocotyl-root transition zone (similar to that in leafy spurge seedlings) located just below the soil surface. A bud forms on the seedling axis at this zone, borne in the axil of a foliarlike organ, the *cataphyll.* The bud arises exogenously by dedifferentiation of corticular parenchyma. This bud, as it grows, becomes a lateral branch arising from the seedling axis. The branch elongates and gradually turns toward the soil surface. Its lower portion becomes the rhizome and the upper portion emerges from the soil and becomes the aerial shoot. In Nebraska, the first rhizomes on western ironweed seedlings are fully formed by mid-June.

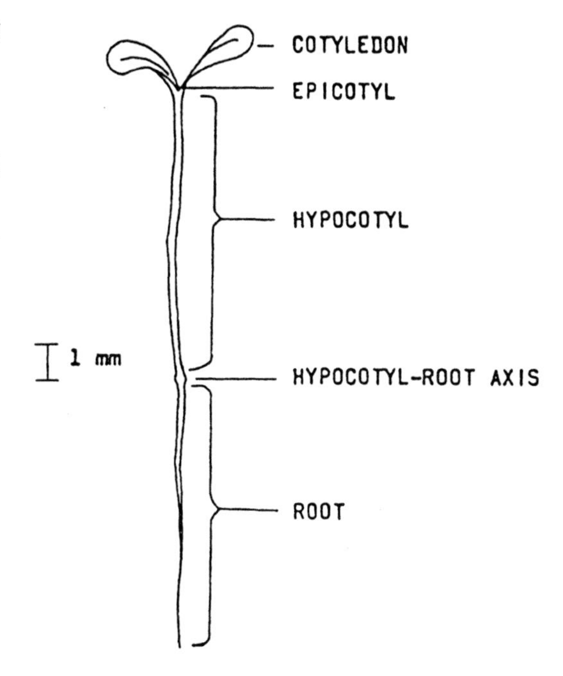

Figure 20.4. Diagram of a western ironweed seedling 6 days after germination. A cortical collar at the hypocotyl-root axis gives rise to the first lateral buds.

Source: Scifres, C.J., and M.K. McCarty. 1969. Seedling development and bud ontogeny in western ironweed. *Weed Science* 17: 83-86.

Bud Dormancy

Mature buds of western ironweed rhizomes have an annual cycle in which each bud is internally dormant and nondormant each year, with transition periods between. Most notable is the annual cycle of high activity (no dormancy) in early spring (March, April, May) and low activity (internal dormancy) in late fall and early winter (September to mid-December), with transition periods between.

The buds sprout in the spring after internal dormancy ends naturally. Dormancy is apparently broken naturally by prolonged exposure to near freezing soil temperatures at a 5-inch (12.7-cm) depth (similar to leafy spurge). However, no bud sprouting occurs until soil temperatures at this soil depth reach 59°F (15°C).

Internal dormancy in western ironweed is a population phenomenon. That is, at any given time during the onset of internal dormancy (July through September), random sampling includes some internally dormant buds and others which are still active and readily sprout. During a single season, western ironweed exhibits periods of correlated inhibition (shoot effect), internal dormancy, immaturity (basal buds),

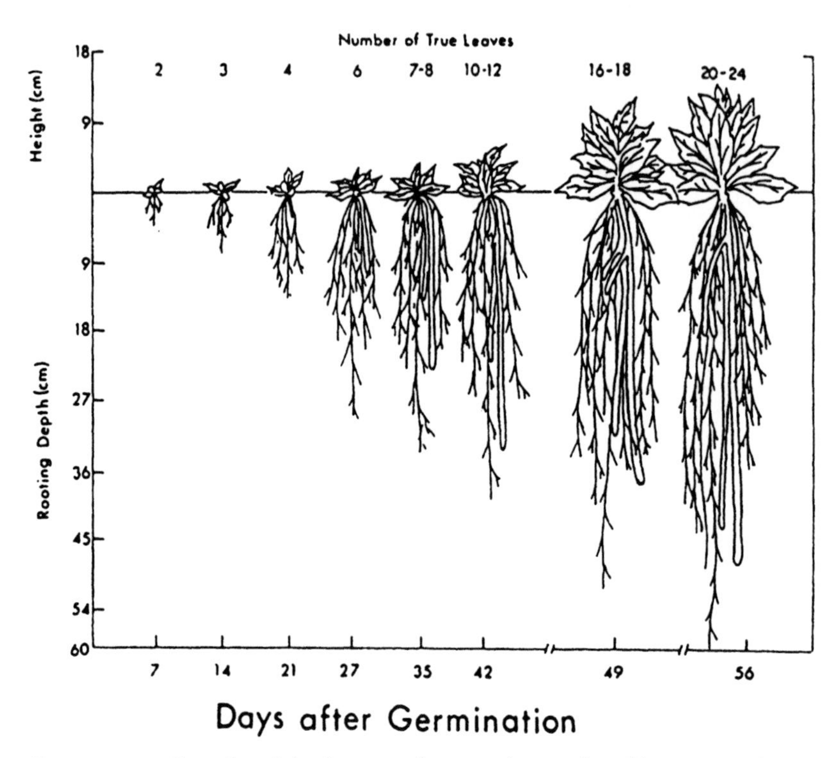

FIGURE 20.5. Growth and development of western ironweed seedlings over an 8-week period under greenhouse conditions.

Source: Scifres, C.J., and M.K. McCarty. 1969. Seedling development and bud ontogeny in western ironweed. *Weed Science* 17: 83-86.

temperature block to sprouting, and apical dominance (shoots and rhizome buds).

Internal dormancy is an important factor in the control of western ironweed. The inability of this perennial weed to make vegetative growth during part of the growing season is of considerable importance in the timing and effectiveness of weed control measures.

REFERENCES

Davis, F.S., and M.K. McCarty. 1966. Effect of several factors on the expression of dormancy in western ironweed. Weeds 14: 62–67.

Linscott, D.L., and M.K. McCarty. 1962. Absorption, translocation, and degradation of 2,4-D in ironweed (*Vernonia baldwinii*). Weeds 10: 65–68.

McCarty, M.K., and C.J. Scifres. 1969. Herbicidal control of western ironweed. Weed Sci. 17: 77–79.

Monson, W.G., and F.S. Davis. 1964. Dormancy in western ironweed and leafy spurge. Weeds 12: 238–239.
Reed, C.F., and R.O. Hughes. 1970. Selected Weeds of the United States. In Agriculture Handbook No. 336. Washington, DC: United States Department of Agriculture, U.S. Government Printing Office, pp. 442–443.
Scifres, C.J., and M.K. McCarty. 1969. Seedling development and bud ontogeny in western ironweed. Weed Sci. 17: 83–86.
Scifres, C.J., and M.K. McCarty. 1969. Vegetative reproduction in western ironweed. Weed Sci. 17: 104–108.

Part Eight

Perennial Broadleaved Weeds Reproducing from Aerial Runners, Stolons, or Creeping Rhizomes

21

Ground Ivy
(*Glechoma hederacea*)

Introduction

Ground ivy (Figure 21.1) is a prostrate, creeping, strongly aromatic, herbaceous, perennial broadleaf weed. It reproduces by seeds and aboveground runners rooting at the nodes, with flowering shoots 4 to 8 inches (10 to 20 cm) high developing at the rooted nodes. Ground ivy is found in damp, fertile soils in lawns, orchards, shaded areas, and waste places.

Distribution

Ground ivy was introduced into the United States from Europe. It is a common weed throughout most of the eastern one-half of the United

FIGURE 21.1. Ground ivy (*Glechoma hederacea* L.). A. Plant habit ×0.5; B. flower cluster ×2.5; C. flower diagram, showing the four ascending stamens ×2.5; D. nutlets ×6.

Source: Reed, C.F., and R.O. Hughes. 1970. Ground ivy (*Glechoma hederacea* L.). In *Selected Weeds of the United States*. Agriculture Handbook No. 366, p. 313. Washington, DC: U.S. Department of Agriculture, U.S. Printing Office.

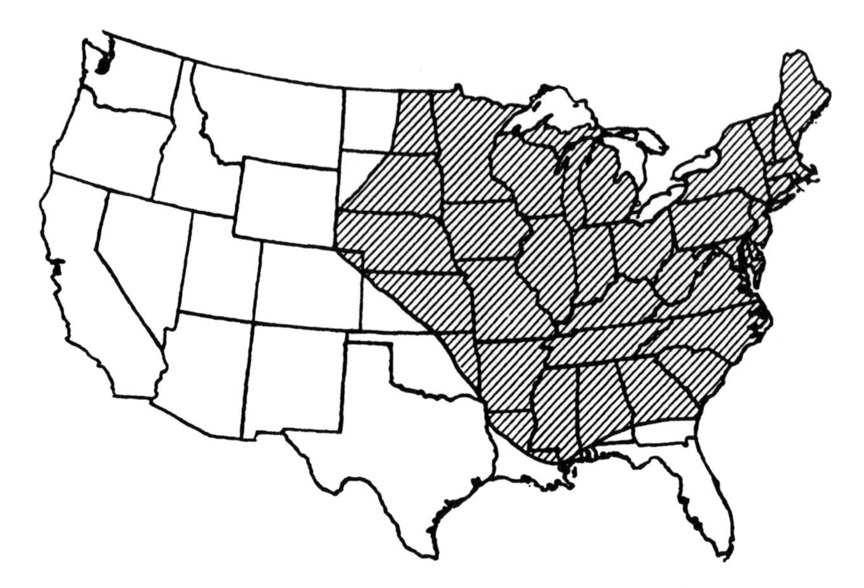

FIGURE 21.2. Distribution of ground ivy in the United states.

Source: Reed, C.F., and R.O. Hughes. 1970. Ground ivy (*Glechoma hederacea* L.). In *Selected Weeds of the United States*. Agriculture Handbook No. 366, p. 312. Washington, DC: U.S. Department of Agriculture, U.S. Government Printing Office.

States, except for parts of some southeast and south central states (Figure 21.2). In Canada, ground ivy is found in the eastern provinces, from Newfoundland to Ontario.

DISTINGUISHING CHARACTERISTICS

Ground ivy is a prostrate, creeping, perennial weed. It reproduces by seeds and aboveground creeping stems (aerial runners). Its fibrous, feeder roots are shallow. Flowering shoots develop at the rooted nodes of the aerial runners. The stems are glabrous or nearly so, creeping or trailing, four-sided (square), 8 to 24 inches (20 to 60 cm) long, rooting at the nodes, with numerous erect, flowering branches.

The leaves are opposite, palmately veined, petioled, rounded or kidney-shaped, 0.5 to 1.5 inches (1 to 4 cm) wide, with scalloped edges. They are glabrous, bright green on both sides, and have a minty odor.

Ground ivy flowers from April to June. The flowers are small and in clusters in the leaf axils. The calyx is tubular (funnel-shaped), two-lipped, bluish-purple to purplish, about 0.5 to 0.7 inch (1.3 to 1.8 cm) long, pubescent, and persistent. The upper lip is two-cleft and the lower lip is three-lobed. The fruit separates into four mottled brown nutlets (seeds), about 2 mm long, dark brown with a small whitish hilum at the base. A *nutlet* is a one-seeded portion of a fruit that fragments at maturity.

MISTAKEN IDENTITY

As a lawn weed, ground ivy may be mistaken for henbit, *Lamium amplexicaule* (Figure 21.3), by the casual observer. Henbit is a herbaceous, biennial or winter annual broadleaf weed with a fibrous root system. Perhaps the most prominent identifying traits are henbit's sessile upper leaves clasping the stem and decumbent, four-sided (square) stems with numerous ascending branches. In contrast, the leaves of ground ivy are petioled and the stems are long, creeping or trailing. Henbit is more widely distributed in the United States than is ground ivy (Figure 21.4).

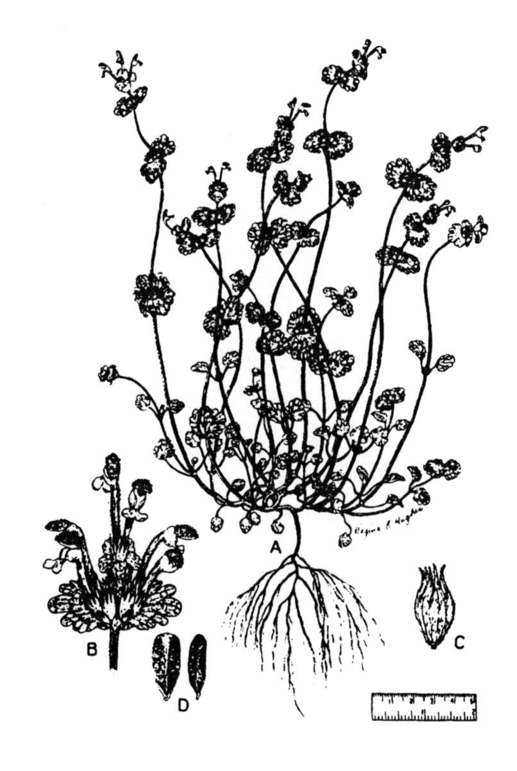

FIGURE 21.3. Henbit (*Lamium amplexicaule* L.). A. Plant habit ×0.5; B. flower clusters showing very short upper internodes ×1.5; C. calyx surrounding nutlets ×4; D. nutlets ×7.5; Note ascending stem branches and sessile upper leaves clasping stems.

Source: Reed, C.F., and R.O. Hughes. 1970. Henbit (*Lamium amplexicaule* L.). In *Selected Weeds of the United States*. Agriculture Handbook No. 366, p. 315. Washington, DC: U.S. Department of Agriculture, U.S. Government Printing Office.

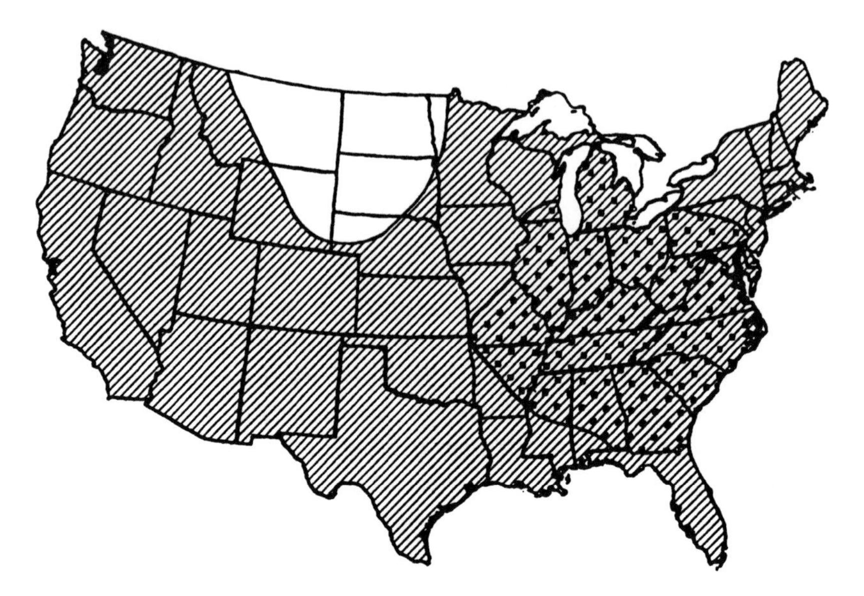

FIGURE 21.4. Distribution of henbit in the United States. Denser cross-hatching denotes infestations of greater economic importance.

Source: Reed, C.F., and R.O. Hughes. 1970. Henbit (*Lamium amplexicaule* L.). In *Selected Weeds of the United States*. Agriculture Handbook No. 366, p. 314. Washington, DC: U.S. Department of Agriculture, U.S. Government Printing Office.

REFERENCES

Korsmo, E. 1954. *Glechoma hederacea* L. In Anatomy of Weeds. Forlag, Norway: Grondahl and Sons, pp. 270–273.

Murphy, T.R. (ed). 1993. Ground ivy (*Glechoma hederacea*). In Weeds of Southern Turf Grasses. Athens: Agricultural Business Office, The University of Georgia, p. 130.

Nelson, E.W., and L. Robison. 1968 (revised). Ground ivy (*Glechoma hederacea*). In Nebraska Weeds. Bulletin No. 101-R. Lincoln: Nebraska Department of Agriculture, p. 38-1.

Reed, C.F., and R.O. Hughes. 1970. Ground ivy (*Glechoma hederacea*). In Selected Weeds of the United States. Agriculture Handbook No. 366. Washington, DC: U.S. Department of Agriculture, Superintendent of Documents, pp. 312–313.

Reed, C.F., and R.O. Hughes. 1970. Henbit (*Lamium amplexicaule* L.). In Selected Weeds of the United States. Agriculture Handbook No. 366. Washington, DC: U.S. Department of Agriculture, Superintendent of Documents, pp. 314–315.

Wax, L.M., R.S. Fawcett, and D. Isely. 1990. Ground ivy (*Glechoma hederacea*). In Weeds of the North Central States. North Central Regional Research Publication No. 281, Bulletin 772. Urbana: College of Agriculture, University of Illinois at Urbana-Champaign, p. 150.

22

Creeping Woodsorrel
(Oxalis corniculata)

Yellow Woodsorrel
(Oxalis stricta)

Introduction

Creeping woodsorrel (Figure 22.1) is a slender, prostrate to semierect, creeping, stoloniferous, cloverlike, herbaceous, perennial broadleaf weed, with new growth ascending up to 8 inches (20 cm). It is one of the most troublesome weeds of greenhouses, nurseries (containers),

Figure 22.1. Creeping woodsorrel, showing prostrate stolon rooting at nodes with flowers and seedpods. a. Flower; b. seedpod; c. two views of seed.

Source: Parker, K.F. 1972. *Arizona Weeds*. Tucson: University of Arizona Press.

lawns, gardens, and flower beds. It grows in dry or moist, usually shaded, soil. Creeping woodsorrel is one of the worst turf weeds in California. It has been cultivated as an ornamental and sometimes sold as "shamrock."

Yellow woodsorrel (Figure 22.2) is an upright, creeping, cloverlike, herbaceous, perennial broadleaf weed, resembling, in general, creeping woodsorrel. Yellow woodsorrel is found in cultivated lands, lawns, roadsides, and woodlands, but the extent of loss from invasion is not known. It has been described as one of the worst weed problems in nurseries.

Distribution

Creeping woodsorrel has become naturalized in the United States; it was introduced from Europe. On a global scale, creeping woodsorrel

Figure 22.2. Yellow woodsorrel (*Oxalis stricta* L.). A. Plant habit ×0.5; B. leaves ×1.25; C. flower diagrams ×2.5; D. seedpods ×1.5; E. seeds ×10.

Source: Reed, C.F., and R.O. Hughes. 1970. Yellow woodsorrel (*Oxalis stricta* L.). In *Selected Weeds of the United States*. Agriculture Handbook No. 366, p. 241. Washington, DC: U.S. Department of Agriculture, U.S. Government Printing Office.

occurs in inhabited regions throughout the world, except for Arctica and Antarctica.

Yellow woodsorrel is found throughout the United States (Figure 22.3), and in Canada, Europe, Africa, Asia, Japan, and New Zealand. In Canada, it is a common weed in the eastern provinces of Prince Edward Island, Nova Scotia, and New Brunswick, and in the extreme southern parts of Quebec, Ontario, and Manitoba. Yellow woodsorrel is by far more common and widespread in Canada than is creeping woodsorrel.

PROPAGATION

Creeping woodsorrel propagates by seeds and by stolons rooting and forming new plants at the nodes.

Yellow woodsorrel reproduces by seeds and by new plants arising from axillary buds at the nodes of rhizomes.

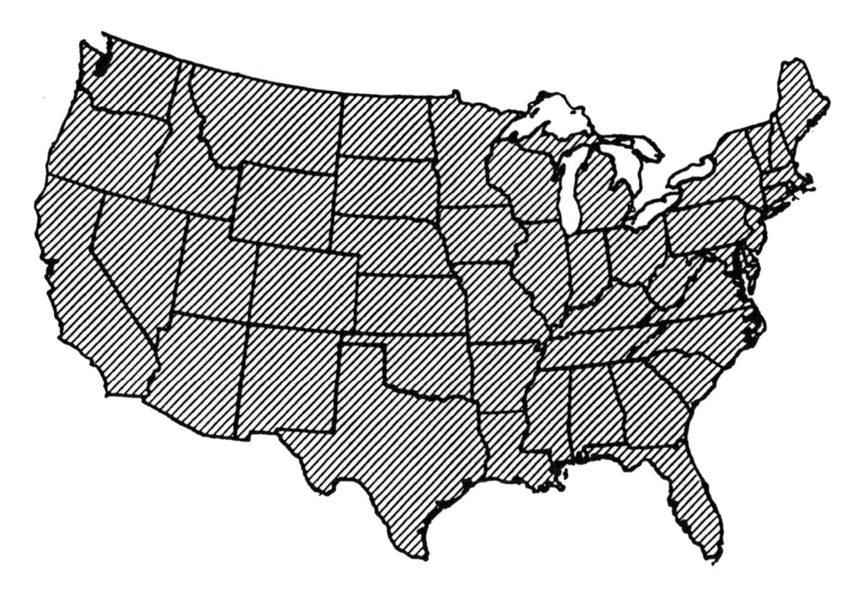

FIGURE 22.3. Distribution of yellow woodsorrel (*Oxalis stricta* L.) in the United States.

Source: Reed, C.F., and R.O. Hughes. 1970. Yellow woodsorrel (*Oxalis stricta* L.). In *Selected Weeds of the United States*. Agriculture Handbook No. 366, p. 240. Washington, DC: U.S. Department of Agriculture, U.S. Government Printing Office.

Spread

Creeping woodsorrel spreads by seeds forcibly ejected from mature seedpods and clonally by creeping stolons, with few to many arising from a taproot. The rooted units may regenerate after weeding or rototilling.

Yellow woodsorrel spreads by seed forcibly ejected from mature seedpods and by creeping rhizomes which form new plants at the nodes.

Perennation

Creeping woodsorrel plants overwinter in warm, protected areas; especially greenhouses. In cold, unprotected areas outdoors, creeping woodsorrel plants grow as summer annuals.

The perennial growth habit of yellow woodsorrel is due to survival of its rhizomes and reproductive axillary buds.

Rhizomes

Plants of creeping woodsorrel do not have rhizomes. In contrast, plants of yellow woodsorrel do have rhizomes.

Distinguishing Characteristics

Creeping Woodsorrel

Creeping woodsorrel is quickly distinguished from yellow woodsorrel by its decumbent growth habit, presence of taproot and stolons, absence of rhizomes, and by its umbellate inflorescence (i.e., the pedicels appear to arise from the same point at the tip of the peduncle), with one to five flowers per peduncle. The pedicels supporting the seedpods are erect to abruptly bent.

Yellow woodsorrel is distinguished from creeping woodsorrel by its erect growth habit, absence of stolons, presence of rhizomes, and by its cymose inflorescence (each floral axis terminating in a single flower and subsequent flowers arising from lateral buds), with one flower per peduncle. The pedicels supporting the seedpods are horizontal to erect.

Creeping woodsorrel is a creeping, decumbent, herbaceous perennial weed. It has a slender taproot, and rhizomes are lacking. It is distinguished by its cloverlike, three heart-shaped leaflets (obcordate with an apical notch) borne at the tip of long petioles. The leaves fold at night. The leaflets are green, purplish, or bronze, and are hairless or thinly haired. The petiole base is auriculate (with auricles). The main stem is determinate in growth. The stolons are 3 to 8 inches (7.5 to 20 cm) long. The lateral shoots borne on the stolons are indeterminate in growth, and they continue to produce flowers as they grow. Seedpods are erect, hairy, cylindrical, ⅓ to 1 inch long, and pointed at the tip. In Arizona, creeping woodsorrel flowers outdoors from February to November.

The flowers of creeping woodsorrel occur in clusters of one to five flowers at the end of slender flower stalks (peduncles) arising from the leaf axils. They have five yellow petals, with or without reddish lines in the throat, 0.14 to 0.5 inch (3.5 to 13 mm) long. There are five overlapping sepals and the five yellow petals are about twice the length of the sepals. Creeping woodsorrel is a facultative self-pollinator, with the closeness of long anthers and stigma or long stigma and mid-length anthers ensuring self-pollination.

The yellowish seedpods of creeping woodsorrel are erect, but their short stalks are sharply bent. The pods are cylindrical, five-angled, hairy, 0.3 to 1 inch (8 to 25 mm) long, pointed at the tip, and contain about 50 seeds per pod (two to many per locule). The tiny reddish-brown seeds are 1.0 to 1.6 mm long, somewhat egg-shaped but flattened with seven to 10 transverse ridges on each face. When the seeds are mature, the pods dehisce by splitting longitudinally along each locule on the side away from the axis (abaxial), forcibly ejecting and scattering an average of 50 seeds up to 13 ft (4 m) from the parent plant.

In greenhouses, creeping woodsorrel is perennial, existing year round as aboveground and belowground plant tissue and/or seeds. In greenhouses with warm night temperatures and moderate light intensity, the plants do not respond to seasons as do those growing outdoors; rather, they complete one life cycle in about 6 weeks from seed to seed, producing many generations of seedlings in a single year. Growing outdoors in cold climates, creeping woodsorrel follows an annual life cycle, overwintering as seeds. However, growing outdoors in areas with mild winters, it follows a perennial lifespan. For the most part, outdoor populations of creeping woodsorrel are established each year from seed.

Yellow Woodsorrel

Yellow woodsorrel is an erect, herbaceous perennial, 2 to 16 inches (5 to 40 cm) tall. The stems are determinate and arise singly from white to pink, succulent, underground perennial rhizomes. Stems are green to purplish, usually erect, unbranched or with several branches from the base, occasionally decumbent but not rooting at the nodes. Leaves are alternate, cloverlike, divided in three deeply cut, heart-shaped lobes (obcordate with an apical notch). The leaves fold at night. The leaflets are 12 to 29 mm wide; petiole base is not auriculate (without auricles). Flowers have five yellow petals 3.5 to 11 mm long, with five sepals. The pedicels supporting the seedpods are horizontal to erect. The seedpods are 8 to 13 mm long, glabrous or pubescent, five locules per pod, and five to 10 or more seeds per locule. The seeds are brown, 1.0 to 1.5 mm long, rarely with light lines on the transverse ridges. The pods dehisce by splitting longitudinally along each locule on the side away from the axis (abaxial), forcibly ejecting and scattering an average of 50 seeds up to 13 feet (4 m) from the parent plant.

Yellow woodsorrel is a facultative self-pollinator, with self-pollination ensured by the closeness of long stigma and anthers or long stigma and mid-length anthers.

In newly cultivated soil, most new plants of yellow woodsorrel arise from seed. In undisturbed ground, yellow woodsorrel regenerates from buds on rhizomes. The leaves of yellow woodsorrel begin to appear aboveground in mid- to late-May in Ontario, Canada, and earlier in warmer climates. Flowering begins shortly afterward and continues throughout the summer.

OXALIC ACID

The leaves of both creeping woodsorrel and yellow woodsorrel contain oxalic acid, and they can be dried and added to salads to impart a "sharp" flavor. However, caution is necessary, as creeping woodsorrel may accumulate lethal concentrations of soluble oxalates.

VARIABILITY

The morphology of both woodsorrel species is very variable, and local forms have been described as new species. Creeping woodsorrel plants vary as to degree of hairiness, plant size, and leaf and flower coloration. Yellow woodsorrel plants vary as to degree of hairiness on leaflets, stems, peduncles (pedicels), and seedpods (capsules), in number of flowers per peduncle, and in length of flower and style. Leaf color is variable in both woodsorrel species and cannot be used as an identifying characteristic.

SEED GERMINATION

Seeds of creeping woodsorrel and yellow woodsorrel require exposure to light for germination; however, only a very low light level is required to stimulate 100% germination. The seeds do not germinate in darkness. The seeds germinate in light at temperatures of 48° to 85°F (8.7° to 29.5°C). Viability of fresh seeds is nearly 100%, regardless of season of production. Seeds stored for as long as 1 year had 83% germination compared to fresh seeds.

REFERENCES

Cudney, D.W. 1985. Oxalis control in turf. Proc. Calif. Weed Conf. 37: 91–94.

DeFelice, M.S., and A.W. Evans. 1985. Weed Identification Guide. Champaign, IL: Southern Weed Science Society, p. 11, OXAST.

Doust, L.L., A. MacKinnon, and J.L. Doust. 1985. Biology of Canadian Weeds. 71. *Oxalis stricta* L., *O. Corniculata* L., *O. dillenii* Jacq ssp. *dillenii*, and *O. dillenii* Jacq ssp. *filipes* (Small) Eiten. Can. J. Plant Sci. 65: 691–709.

Holm, L.G., D.L. Plucknett, J.V. Pancho, and J.P. Herberger. 1977. The World's Worst Weeds. Honolulu: The University Press of Hawaii, pp. 343–346.

Holt, J.S. 1985. A new approach to oxalis control in greenhouse crops. Proc. Calif. Weed Conf. 37: 85–86.

Holt, J.S., and W.J. Chism. 1988. Herbicidal activity of NAA (1-naphthaleneacetic acid) on creeping woodsorrel (*Oxalis corniculata*) in ornamentals. Weed Sci. 36: 227–233.

Parker, K.F. 1972. Arizona Weeds. Tucson: The University of Arizona Press, pp. 194–195.

Reed, C.F., and R.O. Hughes. 1970. Common yellow woodsorrel (*Oxalis stricta*). In Selected Weeds of the United States. Agricultural Handbook No. 336. Washington, DC: U.S. Department of Agriculture, U.S. Government Printing Office, pp. 240–241.

Appendix

Table 1. Root Systems of Simple Perennial Weeds

Fleshy taproots

Common pokeweed (*Phytolacca americana*)
Curly dock (*Rumex crispus*)
Dandelion (*Taraxacum officinale*)

Woody taproots

Chicory (*Cichorium intybus*)
Cinquefoil
  silverweed (*Potentilla argentea*)
  sulfur (*Potentilla recta*)
Hoary vervain (*Verbena stricta*)
Khakiweed (*Alternanthera pungens*)
Trailing four o'clock (*Allionia incarnata*)
Yellow woodsorrel (*Oxalis stricta*)

Tuberous roots

Crazyweed, Lambert (*Oxytropis lambertii*)
Kudzu (*Pueraria lobata*)
Larkspur
  duncecap (*Delphinium occidentale*)
  Geyer (*Delphinium geyeri*)
  tall (*Delphinium barbeyi*)
Waterhemlock, spotted (*Cicuta maculata*)
Woolly loco (*Astragulus mollissimus*)

Fibrous roots (clump or tuft growth)

Broomsedge (*Andropogon virginicus*)
Foxtail barley (*Hordeum jubatum*)
Smutgrass (*Sporobolus yuccifolium*)
Tall fescue (*Festuca arundinacea*)

TABLE 2. Perennial Weeds That Spread by Creeping Roots

Bindweed, field (*Convolvulus arvensis*)
Blue lettuce (*Lactuca pulchella*)
Blueweed, Texas (*Helianthus ciliaris*)
Canada thistle (*Cirsium arvense*)
Groundcherry, clammy (*Physalis heterophylla*)
Hemp dogbane (*Apocynum cannabinum*)
Hoary cress (whitetop) (*Cardaria draba*)
Horsenettle (*Solanum carolinense*)
Milkweed
- broadleaf (*Asclepias latifolia*)
- climbing (*Sarcostemma cyanchoides*)
- common (*Asclepias syriaca*)
- eastern whorled (*Asclepias verticillata*)
- western whorled (*Asclepias subverticillata*)

Poison ivy (*Toxicodendron radicans*)
Red sorrel (*Rumex acetosella*)
Russian knapweed (*Acroptilon repens*; formerly, *Centaurea repens*)
Silverleaf nightshade (*Solanum elaeagnifolium*)
Sowthistle, perennial (*Sonchus arvensis*)
Spurge
- flowering (*Euphorbia corollata*)
- leafy (*Euphorbia esula*)

Stinging nettle (*Urtica dioica*)
St. Johnswort, common (*Hypericum perforatum*)
Texas blueweed (*Helianthus ciliaris*)
Toadflax
- Dalmatian (*Linaria genistifolia* ssp. *dalmatica*)
- yellow (*Linaria vulgaris*)

Trumpetcreeper (Virginia creeper) (*Campsis radicans*)

TABLE 3. Perenial Weeds That Spread by Rhizomes

Creeping rhizomes

Brackenfern (*Pteridium aquilinum*)
Bursage
  skeletonleaf (*Ambrosia tomentosa*)
  woollyleaf (*Ambrosia grayi*)
Canarygrass, reed (*Phylaris arundinacea*)
Cattail, common (*Typha latifolia*)
Goldenrod, Canada (*Solidigo canadensis*)
Hedge bindweed (*Calystegia sepium;* formerly, *Convolvulus sepium*)
Honeysuckle, Japanese (*Lonicera japonica*)
Ironweed, tall (*Vernonia altissima*)
Johnsongrass (*Sorghum halepense*)
Nutsedge
  purple (*Cyperus rotundus*)
  yellow (*Cyperus esculentus*)
Oxeye daisy (*Chrysanthemum leucanthemum*)
Poison oak (*Toxicodendron toxicarium*)
Quackgrass (*Elytrigia repens*)
Snakeroot, white (*Eupatorium rugosum*)
Stinging nettle (*Urtica dioica*)
Woodsorrel, yellow (*Oxalis stricta*)
Yarrow, common (*Achillea millefolium*)

Noncreeping rhizomes

Dallisgrass (*Paspalum dilatatum*)
English daisy (*Bellis perennis*)
Mugwort (*Artemisia vulgaria*)
Plantain
  blackseed (*Plantago rugelli*)
  broadleaf (*Plantago major*)
  buckhorn (*Plantago lanceolata*)
Tall buttercup (*Ranunculus acris*)
Western ironweed (*Vernonia baldwinii*)

TABLE 4. Perennial Weeds That Spread by Means Other Than Roots or Rhizomes

Runners

Buttercup, creeping (*Ranunculus repens*)
Cinquefoil
  common (*Potentilla canadensis*)
  silverweed (*Potentilla anserina*)
Clover, white (*Trifolium repens*)
Ground ivy (*Glechoma hederacea*)
Healall (*Prunella vulgaris*)
Speedwell, common (*Veronica officinalis*)
Watercress (*Nasturtium officinale*)

Stolons

Bermudagrass (*Cynodon dactylon*)
Buttercup, creeping (*Ranunculus repens)*
Woodsorrel, creeping (*Oxalis corniculata*)

Decumbent stems

Chickweed
  field (*Cerastium arvense*)
  mouseear (*Cerastium vulgatum*)
Nimblewill (*Muhlenbergia schreberi*)

TABLE 5. Selected Perennial Weeds Identified by Common and Scientific Names and Grouped as Grasses, Herbaceous Broadleafs, Vines, and Woody Plants

Grasses

Bahiagrass (*Paspalum notatum*)
Bermudagrass (*Cynodon dactylon*)
Broomsedge (*Andropogon virginicus*)
Buffalograss (*Buchloe dactyloides*)
Bulbous bluegrass (*Poa bulbosa*)
Canarygrass, reed (*Phalaris arundinacea*)
Creeping bent (*Agrostis palustris*)
Dallisgrass (*Paspalum dilatatum*)
Desert saltgrass (*Distichlis stricta*)
Foxtail barley (*Hordeum jubatum*)
Giant reed (*Arundo donax*)
Johnsongrass (*Sorghum halepense*)
Knotroot foxtail (*Setaria geniculata*)
Nimblewill (*Muhlenbergia schreberi*)
Orchardgrass (*Dactylis glomerata*)
Porcupinegrass (*Stipa spartea*)
Quackgrass (*Elytrigia repens*; formerly *Agropyrons repens*)
Smutgrass (*Sporobolus indicus*)
Sorghum-almum (*Sorghum almum*)
Squirreltail (*Elymus elymoides*)
Tall fescue (*Festuca arundinacea*)
Tumble windmillgrass (*Chloris verticillata*)
Vaseygrass (*Paspalum urvillei*)
Velvetgrass, common (*Holcus lanatus*)
Velvetgrass, German (*Holcus mollis*)
Wirestem muhly (*Muhlenbergia frondosa*)

Herbaceous broadleaved (dicot)

Alkali sida (*Sida hederacea*)
Bindweed, field (*Convolvulus sepium*)
Bindweed, hedge (*Calystegia sepium*)
Blue lettuce (*Lactuca pulchella*)
Blueweed, Texas (*Helianthus ciliaris*)
Bursage, skeletonleaf (*Ambrosia tomentosa*)

Bursage, slimleaf (*Ambrosia confertifolia*)
Buttercup, tall (*Ranunculus acris*)
Canada thistle (*Cirsium arvense*)
Chickweed, mouseear (*Cerastium vulgatum*)
Chicory (*Cichorium intybus*)
Cinquefoil, silvery (*Potentilla anserina*)
Clover, white (*Trifolium repens*)
Corn spurry (*Spergula arvensis*)
Crazyweed, Lambert (*Oxytropis lambertii*)
Curly dock (*Rumex crispus*)
Dandelion (*Taraxacum officinale*)
English daisy (*Bellis perennis*)
Goldenrod, Canada (*Solidago canadensis*)
Gourd, buffalo (*Cucurbita foetidissium*)
Gourd, fingerleaf (*Cucurbita digitata*)
Ground ivy (*Glechoma hederacea*)
Groundcherry, small (*Chamaesaracha coronopus*)
Groundcherry, smooth (*Physalis subglabrata*)
Healall (*Prunella vulgaris*)
Hemp dogbane (*Apocynum cannabinum*)
Hoary cress (*Cardaria draba*)
  also, whitetop
Hog potato (*Hoffmanseggia densiflora*)
Horehound, white (*Marrubium vulgare*)
Horsenettle (*Solanum carolinense*)
Khakiweed (*Alternanthera pungens*)
Knapweed, diffuse (*Centaurea diffusa*)
Knapweed, Russian (*Acroptilon repens*;
  formerly, *Centaurea repens*)
Larkspur, barestem (*Delphinium scaposum*)
Larkspur, duncecap (*Delphinium occidentale*)
Larkspur, Geyer (*Delphinium geyeri*)
Larkspur, tall (*Delphinium barbeyi*)
Leafy spurge (*Euphorbia esula*)
Locoweed, purple (*Astragulus mollissimus*)
  also, Lambert crazyweed
Locoweed, white (*Oxytropis lambertii*)
Milkweed, broadleaf (*Asclepias latifolia*)
Milkweed, climbing (*Saracostemma cynanchoides*)

Milkweed, common (*Asclepias syriaca*)
Milkweed, eastern whorled (*Asclepias verticillata*)
Milkweed, honeyvine (*Ampelamus albidus*)
Milkweed, western whorled (*Asclepias verticillata*)
Oxeye daisy (*Chrysanthemum leucanthemum*)
Pepperweed, perennial (*Lepidium latifolium*)
Plantain, blackseed (*Plantago rugelii*)
Plantain, broadleaf (*Plantago major*)
Plantain, buckhorn (*Plantago lanceolata*)
Povertyweed (*Iva axillaris*)
Pricklepoppy, bluestem (*Argemone intermedia*)
Ragweed, common (*Ambrosia artemisiifolia*)
Ragweed, western (*Ambrosia psilostachya*)
Red sorrel (*Rumex acetosella*)
Sacred datura (*Datura meteloides*)
St. Johnswort (*Hypericum perforatum*)
Silverleaf nightshade (*Solanum elaeagnifolium*)
Smartweed, swamp (*Polygonum coccineum*)
Snakeroot, white (*Eupatorium rugosum*)
Snakeweed, broom (*Gutierrezia sarothrae*)
Sneezeweed, western (*Helenium hoopesii*)
Sowthistle, perennial (*Sonchus arvensis*)
Spiderling, red (*Boerhaavia coccinea*)
Spiny aster (*Aster spinosus*)
Stinging nettle (*Urtica dioica*)
Trailing four o'clock (*Allionia incarnata*)
Waterhemlock, spotted (*Cicuta maculata*)
Waterhemlock, western (*Cicuta douglasii*)
Western ironweed (*Vernonia baldwinii*)
Whitetop, hairy (*Cardaria pubescens*)
  also, globe-podded whitetop
Whitetop, lens-podded (*Cardaria chalepensis*)
Wild licorice (*Glycyrrhiza lepidota*)
Woodsorrel, creeping (*Oxalis corniculata*)
Yellow toadflax (*Linaria vulgaris*)

Vines

Bindweed, field (*Convolvulus arvensis*)
Bindweed, hedge (*Calystegia sepium*)

Honeysuckle, Japanese (*Lonicera japonica*)
Kudzu (*Pueraria lobata*)
Milkweed, climbing (*Sarcostemma cyanchoides*)
Milkweed, honeyvine (*Ampelamus albidus*)
Morning glory, bigroot (*Ipomoea pandurata*)
Poison ivy (*Toxicodendron radicans*)
Poison oak, Pacific (*Toxicodendron diversilobum*)
Redvine (*Brunnichia ovata*)
Trumpetcreeper (*Campis radicans*)

Woody plants

Blackberry, Alleghany (*Rubus allegheniensis*)
Camelthorn (*Alhagi camelorum*)
Creosotebush (*Larrea tridentata*)
Gambel oak (*Quercus gambelii*)
Gaura, smallflower (*Gaura parviflora*)
Gorse (*Ulex europaeus*)
Mesquite (*Prosopis juliflora*)
Mesquite, velvet (*Prosopis velutina*)
Mesquite, western honey (*Prosopis glandulosa* var. *torreyana*)
Multiflora rose (*Rosa multiflora*)
Poison ivy (*Toxicodendron radicans*)
Poison oak (*Toxicodendron toxicarium*)
Poison oak, Pacific (*Toxicodendron diversilobum*)
Rattlebush (*Sesbania punicea*)
Sagebrush, big (*Artemisia tridentata*)
Saltcedar (*Tamarix pentandra*)
Scotch broom (*Cytisis scoparius*)
Sumac, smooth (*Rhus glabra*)
Whitethorn (*Acacia constricta*)
Youpon (*Ilex vomitoria*)

Glossary

Achene—a dry, indehiscent, one-seeded fruit, formed from a single carpel, and with the seed distinct from the fruit wall.

Acute—ending in a sharp point.

Adventitious—arising or occurring sporadically.

Adventitious buds—buds that occur anywhere on a plant other that at its apices and leaf axils; they may arise from roots, stems, leaves, or other plant parts.

Aerial—aboveground; in the air.

Allelopathic—chemicals produced by plants that affect the interaction between different plants and microorganisms.

Alternate leaves—one leaf attached at each stem node.

Anther—the pollen-producing part of the stamen.

Apex—the tip.

Apical—at the apex.

Apical bud—a group of meristematic cells at the tip of a stem or root; all tissues of the mature axis are ultimately formed from it.

Apical dominance—growth-inhibiting effect of the apical bud or other plant parts, on the axillary buds, or plant parts, located along the stem or rhizome below it.

Appressed—pressed flatly against the surface, soil, or plant part.

Ascending—growing upward in an upturned manner.

Auricle—one of a pair of lateral appendages at the juncture of the sheath and the blade of certain grass leaves.

Axil—the angle between a leaf and a stem in which branches or flowers may arise.

Axillary branch—a branch originating from an axillary bud.

Axillary bud—a bud located in the leaf axil of a stem or rhizome; axillary buds are also called lateral buds.

Blade—the expanded part of a leaf or floral part.

Bract—a small, rudimentary or imperfectly developed leaf that subtends a flower or portion of an inflorescence.

Bristle—a short, coarse, stiff hairlike stem or leaf projection.

Broadleaf or broadleaved plants—dicotyledonous plants with leaf veins radiating, fingerlike, throughout the leaf from a midrib (midvein) and interconnected by a network of finer veins. Broadleaf plants have taproot or fibrous root systems or both.

Bud—the undeveloped or embryonic stage of a stem, root, or flower.

Bulb—an underground, perennial food storage organ consisting of a stem axis and numerous overlapping leaf scales; a bud with fleshy bracts or scales on a short conical stem, usually subterranean, containing stored food reserves.

Calyx—the sepals considered collectively; the outer whorl of sterile floral leaves, consisting of sepals. The calyx is usually green and covers the other flower parts.

Cambium—a layer of meristematic cells that divide mostly in one plane, giving rise to daughter cells from which permanent tissue is ultimately formed.

Capsule—a dehiscent, dry, several-seeded fruit; a dry fruit of two or more carpels, usually dehiscent by valves.

Caraphyll—a foliarlike organ, from which a bud is borne in the axil of a seedling stem.

Carpel—a portion of the ovary or female portion of a flower.

Caudex—a compressed stem at or just below the ground surface; the crown, as at the top of the dandelion taproot.

Clasping—the basal lobes of a leaf blade reaching partly or entirely around a stem.

Collar—junction of the blade and sheath in grass leaves.

Coma—a tuft of hairs attached to the testa of a seed.

Common name—a non-Latin name for a plant species.

Complete flower—a flower possessing all flower parts: sepals, petals, stamens, and pistil(s).

Cordate—heart-shaped; usually used to describe leaves with a pair of rounded basal lobes.

Corolla—the petals considered collectively; the inner set of sterile, usually colored floral parts immediately inside the calyx.

Correlated inhibition—growth inhibition of one plant part due to the influence of another part.

Corymb—a raceme with the lower flower stalks longer than those above, so that all the flowers are at the same level.

Culm—the stem of a grass or sedge.

Cyme—a short and broad inflorescence; a convex or flat flower cluster, the central flowers unfolding first.

Deciduous—leaves falling at maturity or at the end of the growing season.

Decumbent—lying flat, sprawling, prostrate, but with the young growth and tip growing upward.

Dehiscent—referring to dry fruit that splits open at maturity, releasing the seeds.

Dentate—toothed, with outwardly projecting teeth, generally in reference to leaves.

Dicotyledonous plants—having two cotyledons per seed; the leaf veins radiate, fingerlike, throughout the leaf from a midrib (midvein) and are interconnected by a network of finer veins. The plants have taproot or fibrous root systems or both. Dicotyledonous plants are commonly referred to as broadleaf or broadleaved plants.

Diffuse—loosely spreading.

Dioecious—referring to a plant with unisexual flowers, the pistillate and staminate flowers borne by different individuals; a plant with flowers of only one sex.

Elliptic—oval.

Entire—leaf margins without teeth, serrations, or lobes.

Epigeous cotyledons—pulled aboveground by elongating hypocotyl.

Eurasia—designating general areas in both Europe and Asia.

Fibrous—consisting of a mass of fine adventitious roots.

Filament—the anther-bearing stalk of a stamen.

Floret—one of the closely clustered small flowers that make up the flower head of a composite flower. A grass flower consisting of a lemma, palea, stamens and/or pistil.

Flower head—the cluster of small flowers that appear as one flower, as in a dandelion and a daisy.

Follicle—a simple pod opening down the inner suture.

Fruit—the ripened ovary or ovaries with attached parts.

Fusiform—elongated or tapering toward each end.

Glabrous—smooth or hairless.

Glandular hair—a small hair terminating in a small, pinheadlike gland, frequently secreting rosin, wax, or other substances.

Glume—a bract at the base of and enclosing a grass spikelet.

Grass—any plant in the botanical Gramineae family. Grass plants have fibrous roots, stems that are round, hollow (sometimes solid or semisolid), with solid, often swollen, nodes. The leaves are long, narrow, and have parallel veins; they consist of a blade, a sheath, a ligule (usually), and auricles (occasionally). The flowers are inconspicuous, with flower parts mostly in threes or multiples of three, and borne in spikelets. The vascular system is scattered irregularly throughout the stem.

Grasslike—with long and narrow leaves, usually more than ten times longer than broad.

Grass weeds—in general, monocotyledonous, parallel-veined weeds of the botanical family Gramineae.

Head—a dense inflorescence of sessile or nearly sessile flowers.

Herbaceous perennial—a seed-producing perennial plant having succulent, nonwoody, aboveground vegetation that is killed by severe drought, frost, or below-freezing temperatures.

Herbaceous plant—a vascular plant that does not develop persistent woody tissue aboveground.

Herbaceous stems—succulent stems that undergo little or no increase in diameter due to cambium activity.

Hilum—a scar on a seed where it is attached to the fruit.

Hirsute—having rough, bristly hairs.

Hoary—grayish-white, usually due to a covering of fine, white hairs.

Hypogeous cotyledons—cotyledons remain belowground.

Indehiscent—not opening at maturity.

Inflorescence—the flowering part of a plant; the arrangement of flowers on the flowering shoot, as a spike, panicle, head, cyme, umbel, or raceme.

Internal dormancy—internally arrested growth; the inability to grow in a favorable environment.

Internode—the stem between two successive nodes.

Involucre—a circle of bracts under a flower cluster; any leaflike structure protecting the reproducing structure, as in flower heads of Euphorbiaceae.

Lanceolate—lance-shaped; flattened, two or three times longer than broad, widest in the middle and tapering to a pointed apex.

Lateral bud—a bud located in the leaf axil of a stem or rhizome; lateral buds are also called axillary buds.

Leaf sheath—the lower part of a leaf that encircles the stem, as in grasses.

Leaflet—one small blade of a compound leaf.

Ligule—a thin membranous outgrowth or row of hairs at the junction of the leaf sheath and leaf base of most grasses.

Lobe—segment of a simple leaf cut rather deeply into curved or angular segments.

Membranous—thin and transparent.

Midrib—the main or central rib or vein of a leaf or leaflet; midvein.

Monocotyledonous plants—plants with one cotyledon per seed; the plants have long narrow leaves with parallel veins, a vascular system scattered irregularly throughout the stem, a fibrous root system, and flower parts mostly in threes or multiples of three. The majority of monocotyledonous plants are grasses.

Monoecious—both sexes on the same plant or in the same flower.

Node—the part of a stem where one or more leaves are attached, or from which secondary branches emerge.

Nutlet—a one-seeded portion of a fruit that fragments at maturity.

Oblong—elliptical, blunt at each end, having nearly parallel sides, two to four times longer than broad.

Obtuse—blunt or rounded, widest above the middle.

Opposite leaves—leaves attached in pairs, opposite each other, at the same stem node.

Oval or **ovate**—egg-shaped.

Ovary—the seed-bearing part of the pistil.

Palmate—a characteristic of leaflets or lobes that originate from a common point and diverge like the fingers of a hand.

Panicle—loose and irregular, more or less spreading, compound flower cluster, with flowers borne on individual stalks, usually of pyramidal form.

Pedicel—stalk of a single flower.

Peduncle—stalk of a flower cluster or of a single flower.

Perennate—living from season to season.

Perennating organs—vegetative reproductive parts (asexual propagules) of perennial plants.

Perennial—a plant that grows for 3 or more years; the part aboveground may die but new shoots arise from underground roots, rhizomes, or crowns.

Perfect flower—both the stamens and the pistil(s) contained in the same flower.

Perianth—the calyx and corolla together.

Persistent plant part—a plant part remaining attached after the growing season.

Petal—one of the modified leaves of the corolla; usually the colorful part of a flower.

Petiole—the stem or stalk of a leaf.

Pistil—the female flower part; the base part is the ovary in which the seeds are formed.

Pistillate—a flower with pistil(s) only; one without stamens.

Prickle—a stiff, sharp-pointed outgrowth from the epidermis, as in *Solanum*.

Primary growth—development and growth (cellular expansion and elongation) from apical, axillary, and adventitious buds.

Procumbent—lying on the ground.

Propagate—to reproduce.

Prostrate—parallel to or lying on the soil surface.

Pubescent—covered with fine, soft hairs.

Raceme—an elongated inflorescence, with flowers arranged along a stem on individual stalks of about equal length; flowers opening from the base upward.

Rachis—the main stem bearing flowers or leaves.

Receptacle—the end of the flower stalk to which the flower is attached.

Reniform—kidney-shaped.

Reticulate—netlike.

Rhizome—a horizontal, elongated underground stem capable of producing shoots above and roots below at the nodes; differing from roots by the presence of nodes, internodes, scalelike leaves, and axillary buds.

Rosette—a basal, circular cluster of leaves, usually appressed or near ground level.

Runner—a chain of internodes, forming new plants at each node, and giving rise to the next internode from an axillary bud in a leaf axil (one of two axillary buds at each node), e.g., strawberry.

Scale—a highly modified, dry leaf, usually for protection.

Scape—a leafless flowering stem arising from the ground, e.g., onions.

Secondary growth—an increase in diameter of roots and stems of dicotyledons and gymnosperms following primary growth due to the formation of additional vascular tissue from the vascular cambium (a unique layer of cells separating adjacent xylem and phloem tissues).

Seedcoat—testa.

Seedhead—a collection of flowers clustered on a main stem.

Sepal—one of the members of the calyx, usually green in color.

Serrate—leaf margin with sharp teeth pointing forward.

Sessile—lacking a petiole or stalk.

Sheath—the lower portion of a grass leaf that encircles the stem.

Simple leaves—a leaf consisting of a blade not divided into leaflets; unbranched, not compound.

Sinuate—a wavy leaf margin with regularly spaced indentations.

Smooth—lacking hairs, divisions, or teeth.

Spatulate—spoon-shaped.
Species—a distinct kind of plant, such as bermudagrass, dandelion, or broadleaf plantain.
Spike—an elongated inflorescence with sessile or nearly sessile flowers.
Spikelet—a small spike; the ultimate flower cluster of the inflorescence of grasses or sedges consisting usually of two glumes and one or more florets.
Spine—a short, thornlike organ.
Stamen—the male part of a flower; the part of the flower that produces pollen grains.
Staminate—a flower with stamens but no pistils; male-flowered, with stamens only.
Stellate—star-shaped.
Stigma—the terminal pollen-receptive part of the pistil.
Stipule—bractlike appendages at the base of some leaves.
Stolon—a creeping, prostrate, aboveground, horizontal stem that forms a chain of internodes, roots at the nodes, and forms the next internode from the terminal bud at the last node in the chain, e.g., bermudagrass.
Style—the stalklike part of the ovary that bears the stigma.
Succulent—fleshy.
Taproot—a root system with a prominent main root, bearing smaller lateral roots.
Terminal—at the top or the end.
Testa—the seed coat.
Tuber—an enlarged portion of a rhizome or stolon, commonly its terminal end.
Tufted—in compact clusters.
Ubiquitous—found everywhere; in all types of habitats.
Umbel—a flat or rounded flower cluster in which the stalks radiate from the same point, like the ribs of an umbrella; an inflorescence terminating a flower stem, with pedicels arising from a common point of attachment.
Undulate—wavy, as the margins of leaves.
Vascular bundle—a strand of conducting tissue, consisting of xylem and phloem, sometimes separated by cambium.
Vascular cambium—the cambium arising in vascular bundles and giving rise to secondary xylem and phloem.
Veins—the vascular portions of leaves or flowers.
Viable—containing seeds capable of germinating; alive.
Viscid—sticky.
Weed—any plant growing where it is not wanted.
Whorled—three or more structures at a node, as leaves, branches, or floral parts.
Woody plants—stems characterized by secondary growth and an increase in stem diameter due to new cells laid down by the activity of the vascular and corky cambiums during the 1st, 2nd, and later years of growth. Woody plants develop woody tissue (hard, fibrous lignins beneath the bark in trees and shrubs).

Index

Printed in the United States
133184LV00003B/176/P

9 780813 825205